# Ultraclean Transportation Fuels

ACS SYMPOSIUM SERIES **959**

# Ultraclean Transportation Fuels

**Olayinka I. Ogunsola,** Editor
*U.S. Department of Energy*

**Isaac K. Gamwo,** Editor
*U.S. Department of Energy*

**Sponsored by the**
**ACS Division of Fuel Chemistry**

American Chemical Society, Washington, DC

**Library of Congress Cataloging-in-Publication Data**

Ultraclean transportation fuels / Olayinka I. Ogunsola, editor, Isaac K. Gamwo, editor ; sponsored by the ACS Division of Fuel Chemistry.

p. cm.—(ACS symposium series ; 959)

Includes bibliographical references and index.

ISBN 13: 978–0–8412–7409–9 (alk. paper)

1. Synthetic fuels—Congresses. 2. Hydrocarbons—Congresses.

I. Ogunsola, Olayinka I., 1950- II. Gamwo, Isaac K., 1957- III. American Chemical Society. Division of Fuel Chemistry.

TP360.U44 2007
662′.66—dc22

2006052612

The paper used in this publication meets the minimum requirements of American National Standard for Information Sciences—Permanence of Paper for Printed Library Materials, ANSI Z39.48–1984.



Distributed by Oxford University Press



PRINTED IN THE UNITED STATES OF AMERICA

# Foreword

The ACS Symposium Series was first published in 1974 to provide a mechanism for publishing symposia quickly in book form. The purpose of the series is to publish timely, comprehensive books developed from ACS sponsored symposia based on current scientific research. Occasionally, books are developed from symposia sponsored by other organizations when the topic is of keen interest to the chemistry audience.

Before agreeing to publish a book, the proposed table of contents is reviewed for appropriate and comprehensive coverage and for interest to the audience. Some papers may be excluded to better focus the book; others may be added to provide comprehensiveness. When appropriate, overview or introductory chapters are added. Drafts of chapters are peer-reviewed prior to final acceptance or rejection, and manuscripts are prepared in camera-ready format.

As a rule, only original research papers and original review papers are included in the volumes. Verbatim reproductions of previously published papers are not accepted.

**ACS Books Department**

# Contents

## Production from Natural Gas



## Advances in Catalysts



## Upgrading

## Utilization



## Process Modeling and Optimization



## Indexes

# Preface

Transportation fuels account for a significant portion of crude oil consumed in the United States. Hence, the importance of clean transportation fuels in the nation's economic development, environmental and public health improvement cannot be overly emphasized. Of the total manmade air emissions in the United States, highway transportation is responsible for 57% of the carbon monoxide, 30% of the nitrogen oxides, and 27% of the volatile organic compounds emissions. Driven by the continuing desire of the public for personal mobility, dramatic increases in vehicle numbers and miles have been forecasted. This growth will lead to poorer air quality unless ultraclean transportation fuels are developed and deployed.

To address air pollution problems, the U.S. Environmental Protection Agency (EPA) proposed and developed new vehicle emissions standards. Ultralow sulfur standards for diesel fuel began in 2006. Gasoline sulfur standards have been phased-in since 2004. The EPA Tier 2 tailpipe emissions regulations for light duty vehicles have already taken effect. Standards have also been developed for heavy-duty vehicles that would be phased in from 2007 to 2010.

During the next several years, the United States will implement new, stricter Federal and state emission standards for highway vehicles; encounter more volatile global energy markets; face increased cost of technologies for clean transportation vehicle and fuels production; and encounter increased threat of global climate change. Therefore, it is essential to aggressively pursue the research and development of advanced technologies for high efficiency, low-emissions highway vehicles, as well as for the production of ultraclean fuels required for their operation.

The Department of Energy (DOE) utilized the unique scientific capabilities of its national laboratories and partnering with industry to identify, study, and develop advanced fuels for tomorrow's vehicles. Through these efforts, an Ultraclean Transportation Fuels (UCTF) Program was initiated at the beginning of this century.

The DOE UCTF Program was designed to promote partnerships with industry as well as to develop, demonstrate, and deploy technologies and systems to ensure that the nation's future transportation fuels, utilized in advanced highway vehicles, will contribute to dramatically improved environmental quality, energy security, and economic competitiveness. This new generation of advanced fuels will achieve decreases in air pollution, while increasing or maintaining vehicle performance.

In recent years, research and development activities on ultra clean transportation fuels have significantly increased to meet the increasingly demand for clean fuels that will meet the strict environmental regulations.

This book is the result of two symposia on ultraclean transportation fuels presented at the 2001 and 2005 Fall National Meetings of the American Chemical Society. Nineteen papers were presented in three sessions at the 2001 Meeting and fifteen papers were presented at the 2005 Meeting in two sessions. For a more comprehensive product, authors of papers presented at both National Meetings were invited to submit manuscripts for consideration for publication in the book. The manuscripts were peer reviewed.

This symposium series book describes the recent advances in various aspects (production, processing, upgrading, and utilization) of such recent research and development efforts related to ultraclean transportation fuels from a variety of hydrocarbon feedstocks (coal, petroleum coke, biomass, waste oil, and natural gas).

The form of this book is organized in the following six parts:

- general overview
- basic fundamental and applied research of production from various feed-stocks
- novel methods of upgrading for meeting strict environmental standards and fuel specifications
- utilization in environmentally sound ways
- advances in development of new catalysts for more efficient and cost-effective processes
- computational fluid dynamics modeling to enhance reactor design and optimize transportation fuels reactors.

The authors of each chapter are sharing their timely research and their expert insights in this book. The diversity of their experiences is taking the ultraclean transportation fuels advancement forward.

**Olayinka I. Ogunsola, Ph.D.**
Office of Oil and Natural Gas
U.S. Department of Energy
1000 Independence Avenue, SW
Washington, DC 20585
202–586–6743 (telephone)
202–586–6221 (fax)
Olayinka.Ogunsola@hq.doe.gov (email)

**Isaac K. Gamwo, Ph.D., P.E.**
National Energy Technology Laboratory
U.S. Department of Energy
626 Cochrans Mill Road
Pittsburgh, PA 15236–0940
412–386–6537 (telephone)
412–386–5920 (fax)
gamwo@netl.doe.gov (email)

# Overview

## Chapter 1

# Overview of Fundamentals of Synthetic Ultraclean Transportation Fuel Production

**Murray Silk[1], Mark Ackiewicz[1], John Anderson[1,*], and Olayinka Ogunsola[2]**

**[1]Technology and Management Services, 18757 Fredrick Road, Gaithersburg, MD 20879**
**[2]Office of Oil and Natural Gas, U.S. Department of Energy, 1000 Independence Avenue, S.W., Washington, DC 20585**

With ever increasing requirements for clean transportation fuels and liquid hydrocarbon supplies, there is an opportunity to produce significant quantities of synthetic ultra-clean fuels that are essentially sulfur-free. These synthetic fuels can be produced from natural gas, coal, petroleum coke, biomass, and other non-traditional hydrocarbon sources. Most of these products are fungible and compatible with current products and distribution infrastructure and can be produced at costs competitive with conventional crude oil-derived products under certain market conditions.

## Introduction

Currently, petroleum-derived products meet the majority of the world's transportation fuel needs. However, with rising prices and concern over the amount of conventional oil remaining in the world, there is growing interest in producing transportation fuels from alternative energy sources, such as natural gas, coal, petroleum coke, and biomass. Production of synthetic ultra-clean transportation fuels from these diverse feedstocks can supplement world fuel supplies, mitigating our dependence on traditional crude oil for fuels. In addition, these synthetic products are exceptionally low in sulfur and other undesirable components, meeting or exceeding all proposed standards for clean fuels. Ultra-clean synthetic fuels of interest include Fischer-Tropsch liquids, methanol, dimethyl ether and hydrogen.

 

Production of synthetic liquid fuels has a long history with significant advances being made since the initial use of the processes. In the 1920s, two German scientists working at the Kaiser Wilhelm Institute, Franz Fischer and Hans Tropsch, passed coal-derived synthesis gas over metallic catalysts and produced pure hydrocarbons. The Fischer-Tropsch (F-T) process, invented in petroleum-poor Germany, was used by Germany during World War II to produce alternative fuels. Production reached more than 124,000 barrels per day (b/d) in 1944 (*1*).

This process was first used commercially in the 1950s by the South African Synthetic Oil Corporation (Sasol) to produce transportation fuels (gasoline and diesel) using synthesis gas produced by the gasification of coal. The objective was to convert low-grade coal into petroleum products and chemical feedstocks (*2*). Today, Sasol produces the equivalent of 150,000 barrels per day of fuels and petrochemicals from coal via its indirect liquefaction process.

The Fischer-Tropsch process is the most notable of the various synthetic fuel processes and has products that range from lighter hydrocarbon liquids like naphtha to diesel fuel and heavier hydrocarbons like lubes and waxes. The final product mix is dependent upon operating conditions and process units, such as inclusion of hydrocracking functionality. By varying the design and operating conditions in the reaction section, the plant can optimize production of the desired product slate (*3*).

Using non-traditional feedstocks to produce synthetic ultra-clean fuels, including F-T liquids, methanol, dimethyl ether, and hydrogen has numerous benefits including (*4*):

- Expansion of liquid fuel supply.
- Reduction of U.S. dependency on imported oil.
- Enabling of exhaust aftertreatment systems to be used on gasoline and diesel engines that will significantly reduce emissions of nitrogen oxides, carbon monoxide, hydrocarbons, and particulates.
- Production of a sequestration-ready stream of carbon dioxide, thus enabling potential reductions of greenhouse gas emissions.
- Production of fuel which could help reduce world oil price.

## Production of Synthetic Ultra-Clean Fuels

Synthetic ultra-clean fuels are produced from simple chemical building blocks derived from the breakdown of natural gas, coal, petroleum coke, or biomass. Unlike traditional fuel products, which naturally have substantial impurities such as sulfur and aromatic material which must be removed, the synthetic fuels are produced by building up from molecules which are free of impurities. Therefore, the resultant products are essentially pure, requiring minimal or no further treatment.

The production of ultra-clean synthetic fuels from various feedstocks can be thought of as a three-step process. First, synthesis gas (syngas) is produced from the hydrocarbon feed. Second, the syngas, primarily a mixture of carbon monoxide (CO) and hydrogen ($H_2$), can be further processed through a chemical combination and conversion process to manufacture a heavier liquid hydrocarbon and the desired fuel. The most notable is the F-T process, but other processes also exist. Alternatively, if the desired product is hydrogen, the second step is a shift reaction to convert the carbon monoxide to additional hydrogen and carbon dioxide by adding steam. Finally, the desired products must be separated and purified as necessary to meet specifications and to move the products to consumer markets.

## Syngas Production

There are various feedstocks and chemical conversion reaction schemes that can be used to produce synthetic ultra-clean fuels. All start with the production of syngas, a hydrogen/carbon monoxide mixture.

### *Syngas From Heavy Hydrocarbons*

When starting with coal, petroleum coke, biomass or other heavy solid hydrocarbons, the feedstock is gasified by mixing it with oxygen or air in the presence of steam which partially oxidizes the carbon in the feedstock to produce a mixture of CO, $H_2$, and small amounts of other chemicals. The amount of air or oxygen used during gasification is carefully controlled to be less than stoichiometric quantities so that complete combustion does not occur and only a relatively small portion of the fuel burns completely. The reactions take place at high temperatures (1,000°C or more), which is a result of the exothermic combustion reactions (Reactions 1 to 3) required to drive the endothermic reduction reactions (Reactions 4 and 5). Carbon dioxide ($CO_2$), sulfur, nitrogenous compounds, and other impurities are generally present, which must be subsequently removed from the gaseous stream before further processing.

Oxidation (*2*):

$$C + \tfrac{1}{2}O_2 \longrightarrow CO \qquad \Delta H = -123\ \text{kJ/mol} \qquad (1)$$

$$C + O_2 \longrightarrow CO_2 \qquad \Delta H = -406\ \text{kJ/mol} \qquad (2)$$

$$H_2 + \frac{1}{2}O_2 \longrightarrow H_2O \qquad \Delta H = -248\ kJ/mol \qquad (3)$$

Reduction:

$$C + CO_2 \longrightarrow 2CO \qquad \Delta H = 160\ kJ/mol \qquad (4)$$

$$C + H_2O \longrightarrow CO + H_2 \qquad \Delta H = 119\ kJ/mol \qquad (5)$$

The $H_2/CO$ ratio is important to downstream processes and predominately determined by the feedstock, design, and operating conditions of the gasifier. The water-gas shift reaction (discussed later) rapidly comes to equilibrium in a gasifier and is therefore not controlling.

Minerals in the feedstock separate and leave the bottom of the gasifier as an inert slag or ash. Some of the ash is entrained and requires removal downstream in particulate control equipment, such as filtration and water scrubbers. Other trace elements associated with the feedstock are released during gasification and settle in the waste ash or in gaseous emissions.

Competing reactions taking place in the reactor make it imperative to properly control reaction conditions and to remove undesired products and impurities from the reactor environment. Some competing reactions are:

Methane formation:

$$C + 2H_2 \longrightarrow CH_4 \qquad \Delta H = -87\ kJ/mol \qquad (6)$$

$$CO + 3H_2 \longrightarrow CH_4 + H_2O \qquad \Delta H = -206\ kJ/mol \qquad (7)$$

$$3C + 2H_2O \longrightarrow CH_4 + 2CO \qquad \Delta H = 182\ kJ/mol \qquad (8)$$

*Syngas From Natural Gas*

Natural gas is reacted in a similar manner as gasification of solid hydrocarbons. Methane, the primary component of natural gas, is reacted with steam over a nickel-on-alumina catalyst at 750-800°C to produce synthesis gas (Reaction 9).

$$CH_4 + H_2O \longrightarrow CO + 3H_2 \qquad \Delta H = 206\ kJ/mol \qquad (9)$$

Because the steam methane reforming (SMR) process is highly endothermic, it is necessary to supply heat for the process from an external source. Typically, additional natural gas and tail gas from the SMR process are combusted for this purpose. However, opportunities exist for other fuels to supply the heat, such as a facility in Japan that uses heat from a nuclear plant.

Methane in natural gas can also be partially oxidized with oxygen or air to produce synthesis gas (Reaction 10) (*5*). Oxygen is carefully maintained at the appropriate synthesis gas ratio while the reaction is run at high temperature and short contact times. The gas ratio is important to maximize the yield of CO and $H_2$ while maintaining an acceptable level of $CO_2$. Extremely high synthesis gas yields are possible because of the very short residence time in the reactor. Very short reaction times are critical to stop the series reaction at the desired syngas products. The synthesis gas is formed on platinum or rhodium catalysts while simultaneously inhibiting combustion products and coke. Microreactors may be well suited to this process scheme because of the very short required residence time and rapid thermal quenching.

$$CH_4 + \frac{1}{2}O_2 \longrightarrow CO + 2H_2 \qquad \Delta H = -35.7 \text{ kJ/mol} \qquad (10)$$

Natural gas can also be converted into syngas by combining reactions 9 and 10 in one process. Called autothermal reforming (ATR), this process reacts natural gas with both oxygen/air and steam. The reaction is mildly exothermic, which eliminates the need for an external heat source. One drawback of the process is that, if oxygen is used, it requires the construction of an air separation unit. However, extensive research efforts are underway to combine air separation and reforming in one process step (*6*).

## Syngas Conversion

After producing the syngas and purifying by removal of undesired compounds, it can now be converted into a variety of liquid fuels, chemicals, or hydrogen. The F-T process is the most well-known process for chemical conversion and manufacturing liquids. However, other synthetic ultra-clean fuels can be produced, such as methanol and dimethyl ether (DME), from other processes. Alternatively, hydrogen can be separated and purified from syngas and used in engines or, in the future, fuel cells, if they become commercially available.

### *Water-Gas Shift Reaction – Hydrogen Production*

The water-gas shift (WGS) reaction (Reaction 11) is frequently used to produce additional hydrogen from the syngas mixture. This process could be used to adjust the $H_2/CO$ ratio for downstream fuel processes or to produce large volumes of hydrogen to be used as an ultra-clean transportation fuel. Simply separating and purifying hydrogen from the syngas can be done but using the WGS process would more efficiently maximize $H_2$ production and facilitate separation. Using hydrogen as a fuel has received increased emphasis in recent years (*7*).

Water-Gas Shift Reaction:

$$CO + H_2O \longrightarrow H_2 + CO_2 \qquad \Delta H = -40 \text{ kJ/mol} \qquad (11)$$

The WGS process is thermodynamically limited and favored at lower temperatures (*8*). Higher temperatures improve the rate of reaction but lower the yield of hydrogen. Typically, two stages are employed, consisting of a high-temperature shift at 350°C and a low-temperature shift at 190-210°C. Plants use high and low temperature shift catalysts in series, or a single stage with high or medium temperature shift catalyst. High-temperature shift catalysts are typically composed of iron/chromium and low-temperature shift catalysts are typically copper/zinc (Cu/Zn).

Currently, most WGS research efforts have centered on improving the performance of the catalyst and improving the properties of the reactor tube material so that it can withstand higher thermal stresses, increasing heat fluxes and efficiency. Improvements in the process have been proposed in:

- Changing the reactor's operation from a fixed bed to a fluidized bed, and
- Changing the heat supply from external firing to direct heating.

*Fisher-Tropsch Process*

The F-T process is a catalyzed chemical reaction in which carbon monoxide and hydrogen are converted into liquid hydrocarbons of various forms. The distribution of products depends on the catalyst and the process operation conditions. The F-T chain-growth process is comparable with a polymerization process resulting in a distribution of chain-lengths of the products. The primary products of the F-T process are liquefied petroleum gas (LPG), naphtha, diesel, lubes, and waxes.

Fisher-Tropsch Synthesis:

$$CO + 2H_2 \longrightarrow -CH_2- + H_2O \qquad \Delta H = -165 \text{ kJ/mol} \qquad (12)$$

In this exothermic reaction (Reaction 12), after adjusting the $H_2/CO$ ratio, mainly aliphatic straight-chain hydrocarbons are formed. Branched hydrocarbons, olefins, and alcohols are formed in minor quantities.

Typical catalysts used are based on iron (Fe) and cobalt (Co). The Co catalysts have the advantage of a higher conversion rate and a longer life. They are, in general, more reactive for hydrogenation and, therefore produce less unsaturated hydrocarbons and alcohols compared to iron catalysts. Iron catalysts have a higher tolerance for sulfur, are cheaper, and produce more olefin products and alcohols (*9,10*). The catalyst is also key to controlling the WGS reaction within the F-T reactor, with Fe catalysts having a higher propensity in favor of the reaction and Co catalysts having little or no WGS reaction.

There are mainly two types of F-T reactors. The vertical fixed tube type has the catalyst in tubes that are cooled externally by pressurized boiling water. The other process uses a slurry reactor in which preheated synthesis gas is fed to the bottom of the reactor and distributed into the slurry consisting of liquid and catalyst particles. As the gas bubbles upwards through the slurry, it is diffused and converted into liquid hydrocarbons by the F-T reaction. The heat generated is removed through the reactor's cooling coils where steam is generated for use in the process.

The products from the F-T reaction typically do not require any further upgrading. However, if lighter products are sought (such as diesel fuel, naphtha, and LPG), mild hydrocracking of the waxes and lubes is required. Temperature also influences the product mix, with lower temperatures favoring the production of waxes and lubes, and higher temperatures favoring production of naphtha and LPG. Production of diesel fuel falls in the middle of these operating conditions. Generally, these synthetic liquid products are fully fungible and compatible with existing liquid fuels and, therefore, can be easily introduced into the current infrastructure and supply system. In particular, F-T diesel fuel is extremely high quality with a cetane number over 70 while petroleum-derived diesel fuel is about 45. The sulfur and aromatic contents are very low, with sulfur approaching zero depending on the precise process used.

F-T diesel, though it can be used in pure form as a fuel, is more likely to be used as a blendstock. It has a lower density which results in a perceived lower fuel efficiency when compared to standard diesel and has relatively poor cold-start properties. However its excellent properties allow it to be used as a premium blendstock to bring other refinery streams into specification range. Very likely, with tight sulfur specifications, the volume of F-T diesel required for diluting would be high, leaving little to be used in pure form.

Because of the similarity of the F-T diesel fuel to a JP-8 type fuel, it could also be used as fuel for military applications (*11*). F-T fuels are very useful in minimizing emissions (*12*). Compared to traditional low sulfur fuels, they emit less $CO_2$, 25 percent to 35 percent less particulate matter, 100 percent reduction in SOx, 5 percent to 10 percent lower NOx, and slightly less fuel burned because of the increased gravimetric energy density (*11,13*).

Table I provides a summary of some of the typical products, their yield, and quality from F-T synthesis.

**Table I. GTL Product Slate**

| *Product* | *Yield, %* | *Quality* |
|---|---|---|
| LPG | 2% - 4% | similar to other LPG |
| Naphtha | 18% - 26% | straight chain paraffinic<br>Near zero sulfur |
| Diesel | 50% - 70% | High Cetane, >70<br>Near zero sulfur<br>Low density<br>Low aromatics, <1% |
| Lubes | <1% - 30% | High quality |
| Wax | <1% - 10% | High quality |

The F-T process is an established technology that has already been applied on a large scale, although its widespread use is hampered by high capital costs, high operation and maintenance costs, and the volatility of crude oil prices. In particular, the use of natural gas as a syngas feedstock only becomes practical when using "stranded gas," i.e., sources of natural gas far from major consumption points which are impractical to exploit with conventional gas pipelines and LNG technology. There are large coal reserves which may increasingly be used as a fuel source and this technology could be used if conventional oil were to become more expensive. Combination of biomass gasification and Fisher-Tropsch synthesis is also a promising route to produce renewable transportation fuels.

### *Methanol*

Methanol is a clean-burning liquid that can be used to power electricity-generating turbines or as a fuel for automobiles and other vehicles. It can also be a feedstock for a variety of chemicals (*14*). Though most methanol is produced from natural gas, it can be produced from syngas derived from other feedstocks. Methanol contains no sulfur and produces very little nitrogen oxide pollutants when burned, making it one of the cleanest combustion fuels.

The traditional methanol reaction process consists of passing the syngas over a pelletized catalyst which converts the CO and $H_2$ mix into methanol and water (*15*). The reaction is very exothermic and careful design of heat control equipment is critical. Cooling limits the efficiencies of the reactors and heat must be carried away from the temperature sensitive catalysts. Several companies and organizations have developed processes to optimize this reaction through innovative engineering design.

A recent successful commercial-scale demonstration in the Eastman Chemical facility in Kingsport, Tennessee, of a liquid phase methanol process is particularly suitable for coal-derived syngas (*15*). Using a liquid phase, the process permitted effective control of heat generated from the highly exothermic reaction. This approach achieved much higher synthesis gas conversion per pass than the gas phase counterpart. With the catalyst being temperature sensitive, diluted syngas is generally used to allow better temperature control. Therefore, the improved heat control by using a liquid phase provides a more effective and efficient process.

### *Dimethyl Ether*

Dimethyl ether (DME) has been used as a fuel and propellant. It is miscible with most organic solvents and is currently under study as a fuel additive for diesel engines due to its high volatility and high cetane number. DME, like other synthetically derived fuels, can be produced from syngas, is essentially sulfur free, contains nearly zero aromatic compounds, and is considered an excellent substitute for conventional diesel and liquefied petroleum gas (LPG). DME has several physical properties similar to those of LPG.

DME is superior to methanol as a fuel in terms of environmental friendliness and chemical toxicity (*16*). It is considered to be an additive to diesel or LPG and not a replacement. Firing DME-diesel blends creates challenges for the engine manufacturer. The fuel must be injected at a high enough pressure to keep the dimethyl ether in liquid form. The specific wear characteristics of engines operating with this fuel remains to be studied (*17*).

There are several technologies available to produce DME. One method uses methanol, which is catalytically dehydrated over a silica alumina catalyst to produce DME in a slightly exothermic reaction (Reaction 13). Subsequently, DME is separated from the other components in the product stream.

$$2CH_3OH \longrightarrow CH_3OCH_3 + H_2O \tag{13}$$

In the temperature range of normal operation, there are no significant side reactions and the conversion rate is high.

## Product Separation and Purification

F-T diesel, DME, methanol, and other synthetic fuels require very little, if any, further processing before entering the normal product distribution systems.

Simple distillation to separate the diesel from naphtha and other products is normally required. Hydrocracking of heavier lubes and waxes to maximize diesel yield from the F-T process may also occur.

When producing hydrogen as the final product, impurities such as CO, sulfur compounds, and other trace contaminants must be removed, particularly for application in fuel cells. Currently, pressure swing adsorption (PSA) is commonly used for the separation and purification of hydrogen from mixed gas streams. PSA systems are based on selective adsorbent beds. The gas mixture is introduced to the bed at an elevated pressure and the solid adsorbent selectively "adsorbs" certain components of the gas mixture, allowing the unadsorbed components, in this case hydrogen, to pass through the bed as purificd gas.

New technologies are being investigated that can improve hydrogen production, separation, and purification from hydrocarbon materials. Membranes, as an example, show promise. Membranes act somewhat like a filter, allowing certain materials to pass through while blocking others. However, there are many different types of membranes and membrane materials being investigated, and the physical and chemical means they employ to separate hydrogen from mixed gas streams can be very different. Some examples of the materials being considered are ceramics, metals (such as palladium), and ceramic-metal composites (i.e., cermets). Some of these membrane technologies also have the capability to combine WGS functionality with separation and purification, potentially eliminating the need for two separate process units.

## Commercialization

Currently, there are over 385 gasifiers in operation around the world at 117 plants that have over 45,000 thermal megawatts ($MW_{th}$) of syngas output capacity (*18*). Coal and petrochemical by-products are the predominant feedstocks, with biomass-based applications producing a very small volume of synthesis gas.

Since the late 1990s nearly every major oil company announced plans to build pilot plants or commercial plants to produce synthetically derived diesel fuel through improved processes (*19*). Stringent diesel exhaust emission standards and fuel specifications are compelling the petroleum industry to revisit these processes to competitively produce aromatic and sulfur level complying diesel fuel. An important part of the success of the processes is the increased reactor capacity proven within the last 10 years using slurry-bed reactors that have 100 times the capacity over some pre-1990 reactors, thus offering lower cost.

Commercial adoption of the technologies to produce synthetic ultra-clean fuels has been minimal due to the historically low prices for and availability of traditional hydrocarbons. However, with the recent increases in world oil price, interest in synthetic fuels has also increased. Construction of commercial facilities has been limited to unique niche opportunities to leverage remote or stranded supplies of hydrocarbons, particularly natural gas.

Three units currently operate in South Africa (Table II), mostly as part of Sasol's operations. A natural gas-to-liquids (GTL) plant built in 1993 and owned by Shell operates in Bintulu, Malaysia and was recently upgraded. A plant is under construction in Qatar and another is in bidding in Nigeria. The Oryx GTL plant in Qatar is nearing completion and should start production in 2006, while the Escravos GTL plant in Nigeria is expected to begin production in 2009. The Oryx plant, named after the national animal of Qatar, will use over 9 million $M^3$/d of natural gas to produce 24,000 barrels per day (b/d) of diesel, 9,000 b/d of naphtha, and 1,000 b/d of LPG (*20*). China is currently nearing completion of a 20,000 b/d plant that will produce liquids from coal and has several other projects under development. Other synthetic fuel projects are also being considered throughout the world in countries such as India, Australia, and New Zealand.

**Table II. Commercial Scale GTL Plants**

| *Company* | *Location* | *Size, (b/d)* | *Comment* |
|---|---|---|---|
| Sasol I | Sasolburg, South Africa | 5,600 | 1955, feed coal |
| Sasol II/III | Secunda, South Africa | 124,000 | 1955/1980<br>Feed coal |
| Petro SA | Mossel Bay, South Africa | 22,500 | 1991, gasoline/diesel<br>Sasol technology |
| Shell MDS | Bintulu, Malaysia | 14,000 | 1993, diesel and chemicals<br>natural gas feed |

Although the U.S. invested heavily in demonstration and pilot projects for coal liquefaction technologies in the last 30 years, no commercial plants have been built in this country. Several demonstration facilities have been built or approved which may lead ultimately to more significant installations. Most recently, a 5,000-barrel per day plant has been approved in Gilberton, Pennsylvania using anthracite waste as a feedstock (*21,22*). The plant will also co-produce 41 MW of electricity for export. This plant will cost $612 million and is partially funded by the Department of Energy and will use Sasol and Shell technology. This project has an added benefit of using waste coal as the feedstock, thus mitigating an existing environmental concern.

## Challenges and Incentives to Commercialization

Various challenges exist to the widespread commercial deployment of synthetic ultra-clean transportation fuels. Market, environmental, technical, and readiness challenges have all played an important role in limiting commercialization.

### Market Challenges

Some consuming and producing countries have come to consider synthetic fuels production as one component of a strategy to diversify energy production and consumption. For example, Qatar is striving to monetize its large gas deposits to maintain revenue streams as their oil production rates are projected to decline. With much of the world's reserves of oil located in the Middle East, consuming countries see using alternate feedstocks as a way to diversify and secure their fuel supply sources.

Also, it is clear that energy demand continues to grow at historic rates, mostly due to rapid growth in the Asia-Pacific region, notably China and India. With this increased demand has come greater volatility and higher world oil prices. While higher oil prices have renewed interest in synthetic fuel production, the possibility of a return to lower prices has limited the commercial deployment of synthetic fuel technologies.

Financial incentives can help reduce the uncertainty and risk of volatile world oil prices and improve commercial deployment opportunities for synthetic fuel production. Options such as price floors, price ceilings, loan guarantees, product off-take agreements, or other incentives could have a positive impact on the ability for industry to move forward with synthetic fuel projects.

### Environmental Challenges

Emission standards for diesel vehicles have tightened or are expected to tighten considerably in many countries and regions. These standards have resulted in emission or content limits on sulfur, particulates, aromatics, and nitrogen oxides. Synthetic ultra-clean transportation fuels are virtually free of sulfur and are primarily paraffinic. Therefore, they are considered a potentially important blending element in satisfying environmental and auto emission requirements.

With growing concerns over climate change caused by manmade emissions, advances and additional investments may be needed to capture and store carbon dioxide produced during synthetic ultra-clean fuels production. To avoid emission of $CO_2$ into the atmosphere, $CO_2$ can be concentrated, captured, and permanently stored. However, capture and storage technologies are still relatively new and additional testing and development are needed to prove this technology.

Water use requirements are an important issue for consideration, particularly for areas with low rainfall or limited water resources. In-situ gasification and the use of air cooling in place of water cooling for the major processes can substantially reduce water requirements.

Land use is another concern. Additional natural gas production, coal mining, and growth of dedicated biomass feedstocks will impact land use, alter topography, or impact ecological systems. Mitigation and reclamation strategies would be required to offset some of these changes.

## Technical Challenges

Studies estimate that there are various synthetic fuel production technologies and process schemes available that can be competitive with petroleum-derived fuels under certain market conditions and prices (*23*). However, the high capital requirements and long planning and construction lead times create substantial risks. Additionally, the limited introduction of some of these technologies poses significant technical risks which may be unacceptable to project developers and investors.

## Readiness Challenges

The large level of construction that would be required to build a new generation of fuel production plants may be faced with a scarcity of engineering and skilled labor for plant design and construction. This problem is already evident in some remote locales where large facilities are progressing such as the large tar sands deposits in Canada. There may likely be a worldwide scarcity of process and critical equipment manufacturers and contractors. Raw material costs, such as the cost of steel, may also increase due to greater demand. Additionally, the availability or construction of infrastructure, such as railways, roads, pipelines, and other support functions may limit or delay implementation of projects.

## References

1. *Early Days of Coal Research*, U.S. Department of Energy, Office of Fossil Energy, http://www.fossil.energy.gov/aboutus/history/syntheticfuels_history.html
2. *Sasol's Unique Position in Syngas Production*; van Dyk, J.C., Keyser, M.J., Coertzen, M.; Sasol, Sasolburg, South Africa, Presented at the 2004 Gasification Technologies Conference.
3. *Stranded Gas, Diesel Needs Push GTL Work*, Rahmim, I. I., Oil and Gas Journal, March 14, 2005.
4. *Coproduction of Ultra Clean Transportation Fuels, Hydrogen, and Electric Power From Coal,* Gray, D., Tomlinson, G. Mitretek Systems, Mitretek Technical Report 2001-43, July 2001.
5. *The Catalytic Partial Oxidation of Methane in a Microchannel Chemical Reactor,* Tonkovich, A., Zilka, J., Powell, M., Call, C.; Pacific Northwest National Laboratory, Richland, WA.
6. *Engineering Development of Ceramic Membrane Reactor Systems for Converting Natural Gas to Hydrogen*, U.S. Department of Energy, Office of Fossil Energy Project Fact Sheet, http://www.fossil.energy.gov/fred/feprograms.jsp?prog=Fuels+-+Hydrogen
7. *State of the Union Address*, President George W. Bush, January 28, 2003. http://www.whitehouse.gov/news/releases/2003/01/20030128-19.html
8. *Hydrogen From Coal Program Research, Development, and Demonstration Plan for 2004 through 2015,* U.S. Department of Energy: Office of Fossil Energy, September 2005.
9. *Fischer-Tropsch Synthesis,* Energy Research Centre of the Netherlands, http://www.ecn.nl/en/bkm/rd-programme/biofuels-and-refinery-processes/fischer-tropsch/fischer-tropsch-synthesis/
10. *Small GTL Facilities Can Make Money,* Yakobson, D., Rentech, Inc. Presentation to CWC Group Conference, London, U.K., May 2003.
11. *OSD/AT&L Clean Fuel Initiative,* Presentation, Barna, T., Assistant Deputy Under Secretary of Defense, 2005. http://www.cffs.uky.edu/C1/2005%20Meeting/WPAFB%20Harrison%20Invited.pdf#search='barna%20AND%20OSD'
12. *Reducing Dependence on Imported Oil: Zero Sulfur Transportation Fuels from Domestic Coal,* Gray, D., Tomlinson, G., Mitretek Systems, NCEP Forum: The Future of Biomass and Transportation Fuels, June, 2003.
13. *Overview of Gas-to-Liquids Program: Its Role in Ultra-Clean Transportation Fuels Initiative and Commercialization Strategy,* Venkataraman, V., Workshop on Alternative Fuels for Ferries and Other Vessels, Alameda, CA, 2000.

14. *Methanol-Producing Clean Coal Technology Project Moves into Construction in Kingsport, Tennessee,* Fossil Energy Techline, October 1995.
15. *Commercial-Scale Demonstration of the Liquid Phase Methanol Process – Project Performance Summary,* Air Products Liquid Phase Conversion Company, June 2004. http://www.netl.doe.gov/technologies/coalpower/cctc/resources/pdfs/lpmeoh/LPMEOH_PPS.pdf
16. *Dimethyl Ether As Alternate Fuel,* SRI Consulting, Process Economics Report 245A, August 2005.
17. *DOE to Expand Penn State Research to Study Diesel Engine Durability Using Future Clean Fuel Blends,* Office of Fossil Energy, DOE Techlines, December 11, 2000.
18. *Current Industry Perspective on Gasification – Robust Growth Forecast,* National Energy Technology Laboratory, Pittsburgh, PA, September 2005.
19. *Synthetic Diesel Fuel,* California Energy Commission, http://www.energy.ca.gov/afvs/synthetic_diesel.html
20. *The Revolutionary World of Synthetic Fuels, 18th World Petroleum Conference,* paper by Foster Wheeler, Haldor Topsoe, Oryx GTL and Sasol Chevron, September 2005.
21. *Senators Santorum and Specter Announce Authorization for Nation's First Clean Coal-to-Liquid Fuel Plant,* July 29, 2005, http://www.ultracleanfuels.com/articles/sensant_072905.htm
22. *Benefits of the Gilberton Coal-to-Clean Fuels and Power Project,* Office of Project Management, National Energy Technology Laboratory, December, December 2003.
23. *Reducing Dependence on Imported Oil: Zero Sulfur Transportation Fuels from Domestic Coal*, Gray, D., Tomlinson, G., Mitretek Systems, NCEP Forum, The Future of Biomass and Transportation Fuels, June 2003.

# Chapter 2

# Ultraclean Transportation Fuels from Coal: An Overview

**John Shen[1], Edward Schmetz[1], Gary J. Stiegel[2], John C. Winslow[2], Robert M. Kornosky[2], Diane R. Madden[2], and Suresh C. Jain[3]**

**[1]U.S. Department of Energy, 19901 Germantown Road, Germantown, MD 20874**
**[2]National Energy Technology Laboratory, U.S. Department of Energy, P.O. Box 10940, Pittsburgh, PA 15236**
**[3]National Energy Technology Laboratory, U.S. Department of Energy, 3610 Collins Ferry Road, Morgantown, WV 26507**

Coal is the most abundant hydrocarbon resource that is in commercial production in the U.S. It can be converted to ultra-clean liquid transportation fuels via a technology known as indirect liquefaction. These fuels, known as CTL (coal-to-liquids) fuels, have similar attributes as GTL (gas-to-liquids) fuels, including zero sulfur and aromatic contents. Early commercial production of CTL fuels in the U.S. is likely to take place in an IGCC (integrated gasification combined cycle) complex under the coproduction (poly-generation) mode to coproduce power, fuels, chemicals, and steam. This paper will review some of the work supported by the DOE (U.S. Department of Energy) to facilitate the commercial implementation of the coproduction strategy. It will also highlight the new initiatives in using CTL fuels as feedstocks for distributed hydrogen production. Finally, a brief comparison will be made between the CTL and GTL diesel fuels.

## Introduction

A recent paper by EIA (Energy Information Administration) within DOE (U.S. Department of Energy) indicates that CTL (coal-to-liquids) fuels, under the High Price B Case, could become a viable supplement to the petroleum liquids supply in U.S. by 2010 (1). CTL fuels made from the indirect liquefaction technology are free of sulfur and aromatics, and also have clean combustion properties. These fuels are compatible with the petroleum-based ultra-clean transportation fuels mandated by EPA (U.S. Environmental Protection Agency) for 2006 and beyond to help meet the new stringent specifications for vehicle emissions. The CTL fuels can be used either as a blending component or as neat fuel.

In indirect coal liquefaction, coal is first gasified to form synthesis gas (mostly carbon monoxide and hydrogen), which is cleaned to remove impurities and then converted to clean liquid fuels over catalysts using a technology known as Fischer-Tropsch (F-T) synthesis. The primary F-T products are then upgraded to specification fuels using conventional refining technologies. CTL plants are known to be capital intensive, with syngas production and cleanup accounting for two-thirds of the cost. DOE has been supporting activities aimed to reduce the capital cost of this technology, including work on advanced coal gasifier, improved syngas cleanup, and new slurry-phase F-T reactors. Significant advances have been made in all areas. A scoping economic study indicates that CTL fuels based on the integrated operation of these advanced technologies could be competitive with petroleum crude in the mid-$30 per barrel under the coproduction mode (2).

In the coproduction mode, CTL is produced along with electric power, high-value chemicals and steam by incorporating an F-T synthesis step into an IGCC (integrated gasification and combined cycle) complex. A schematic diagram of the coproduction mode is shown in Figure 1. This mode is deemed to offer the earliest opportunity for the commercial production of CTL fuels in the U.S. because of the near-commercial readiness of IGCC technology. This paper will review the status of activities supported by DOE to help usher in the early commercial production of CTL fuels.

## Early Entrance Coproduction Plant

In 1999, DOE issued a solicitation seeking proposals to evaluate the technical and financial feasibility of projects known as EECP (early entrance coproduction plant). These plants are aimed at demonstrating the operations of advanced CTL technologies in integrated mode at pre-commercial or commercial scale units. The study was structured in three phases: Phase I (concept definition and research/development and testing planning), Phase II

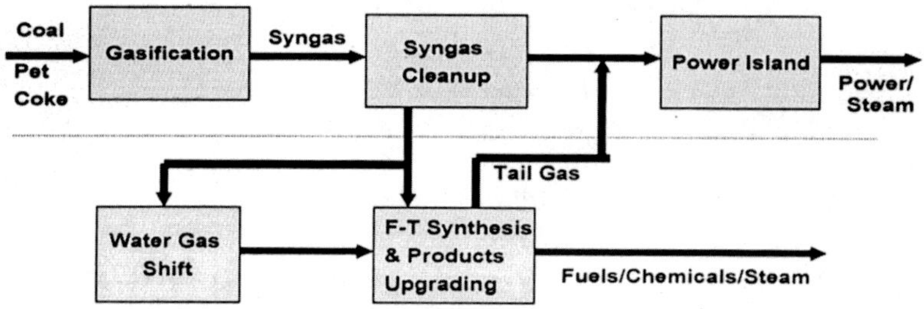


*Figure 1. CTL (coal-to-liquids) fuels production via coal gasification – the co-production mode.*

(research and development and testing), and Phase III (preliminary engineering design). Construction and operation of the EECP are beyond the scope of this solicitation. Successful operation of EECP would mitigate technical risks which are deemed essential in undertaking large-scale commercial CTL projects.

Two projects were awarded under the EECP solicitation. The first one was with WMPI (Waste Management Processors, Inc.) and the other with Texaco Energy Systems LLC. Both projects are aimed at coproducing CTL fuels and electric power from coal and low-cost feedstocks using proprietary iron catalysts. They are to be sited adjacent to existing infrastructures to help reduce the capital costs. Highlights of these two projects are discussed below.

## Waste Management Processors, Inc. (WMPI) Project

Prime contractor: WMPI
Subcontractors: Shell Global Solutions, Uhde GmbH, Sasol Synfuels International, and Nexant, Inc.
Location: Gilberton, Pennsylvania
Feedstock: anthracite coal waste (culm) at 1.4 million tons per year
Products: 5,000 barrels per day of ultra-clean fuels, 39 MW electricity for export, sulfur, and steam.
Key process blocks: Shell gasification complex, Sasol slurry-phase F-T synthesis unit (iron catalyst), ChevronTexaco F-T product workup unit, and combined cycle power plant.

WMPI plans to license the two key technologies of gasification and F-T synthesis from Shell and Sasol which offer guaranteed performance. Syngas from the Shell coal gasifier will first be shifted to increase the hydrogen to carbon monoxide ratio to 1.5 before feeding to the F-T synthesis step. The energy contents in CTL fuels and electric power export have a ratio of 89 to 11. More updated data on the project can be found elsewhere (3)

### Texaco Energy Systems LLC Project

Prime contractor: Texaco Energy Systems LLC
Subcontractors: General Electric (GE), Praxair, and Kellogg Brown and Root (KBR).
Location: Motiva Refinery, Port Arthur, Texas
Feedstock: Petroleum coke at 1,235 short tons per day
Products: 457 barrels per day of wax, 125 barrels per day of F-T diesel, 35 barrels per day of F-T naphtha, 55 MW power for export, sulfur, and steam.
Key process blocks: Texaco gasification complex, Rentech slurry-phase F-T synthesis unit (iron catalyst), Bechtel Hy-Finishing$^{SM}$ F-T product workup unit, and combined cycle power plant.

The Texaco gasification technology is relatively mature and therefore requires minimal additional R&D work. The Rentech slurry-phase F-T technology, on the other hand, is still under development. This new F-T technology will directly process the syngas from Texaco coal gasifier with a hydrogen to carbon monoxide ratio of 0.76, leading to higher plant efficiency. The energy contents in CTL fuels and electric power export have a ratio of 38 to 62. This choice takes into consideration the steady revenues from the off-take of electric power via arrangements through the project host site and the need to obtain scaleup data for the Rentech slurry-phase F-T reactor.

Phase I study showed the project could yield an 11.9% internal rate of return by assuming a zero cost for coke feedstock and selling electric power at market price and F-T wax as premium products. Based on this favorable result, Texaco proceeded to Phase II experimental work which validated the key design parameters used in Phase I study. The Texaco EECP project was completed in 2003 after the completion of Phases I and II (4, 5). Phase III work was cancelled because of the unavailability of the project host site after the Chevron and Texaco merger.

### New Development in EECP Activities

In 2003 WMPI was selected for award negations for commercial coal-based coproduction applications under Round I of the DOE CCPI (clean coal power initiative) solicitation. Its work scope is to construct and operate a commercial CTL plant as designed under the WMPI EECP project mentioned above.

## CTL Fuels for Distributed Hydrogen Production

In 2003, the Hydrogen from Coal program was initiated in the Office of Fossil Energy within DOE as a part of the Hydrogen Fuel Initiative announced

## Chapter 3

# Clean Fuels Production Using Plasma Energy Pyrolysis System

**Ron Patun[1], Jay Ramamurthi[2], Mark Vetter[2], Arthur Hartstein[3], and Olayinka I. Ogunsola[3]**

**[1]National Defense Center for Environmental Excellence (NDCEE), Environmental Technology Center, 100 CTC Drive, Johnstown, PA 15904**
**[2]EnerSol Technologies, Inc., 2722 Merrilee Drive, Suite 310, Fairfax, VA 22031**
**[3]Office of Oil and Natural Gas, U.S. Department of Energy, 1000 Independence Avenue, S.W., Washington, DC 20585**

The PEPS® (Plasma Energy Pyrolysis System) Gasification system, when used in conjunction with existing Fischer-Tropsch technology, can be used to produce ultra-clean liquid fuels from low value, difficult-to-process wastes and by-products. Over 60 million tons of petroleum coke are produced in the U.S. annually as a by-product of petroleum refining. The PEPS® Gasification system from EnerSol Technologies, Inc. provides a novel and effective means to convert low value grades of petroleum coke into high quality synthesis gas. Further processing of the synthesis gas though a Fischer-Tropsch unit will produce ultra-clean liquid fuels. In an alternate configuration, the PEPS® gasification system can be used to produce hydrogen.

## Background – Fuels Needed and an Available Feedstock

The U.S. currently imports about sixty (60) percent of its oil requirements (*1*), which is expected to increase to about 70 percent by the year 2025 (*1*). This reliance on foreign sources of oil has created both national and economic security issues for the U.S. It is desirable to produce liquid transportation fuels from alternative sources. The Fischer-Tropsch (F-T) process can be used to produce liquid fuels from synthesis gas (syngas), a mixture of hydrogen and carbon monoxide. Liquid fuels produced from the F-T process have very low levels of sulfur compared to petroleum products; these ultra-clean fuels are environmentally friendly. However, syngas is commonly produced from natural gas, which has become significantly more expensive in recent years (*2*). Alternative, less expensive feedstocks for syngas production can reduce the costs of liquid fuels produced through the F-T process.

The U.S. produces over 60 million tons per year of petroleum coke (petcoke) at a number of refineries (*3,4*), and the trend indicates higher production in the future. The petcoke is a by-product of thermal cracking of crude oil. Several grades of petcoke are produced. The grade of petcoke produced is affected by the quality of the crude oil, the type of process used to produce the petcoke, and the operating conditions of the process. Although certain grades of petcoke (e.g., "anode grade" petcoke) have economic value, the majority of the petcoke produced is "fuel grade" petcoke. This type of petcoke is characterized by high levels of sulfur and metals, and it has low economic value. Some petcoke is disposed of as a waste by-product by the refiners.

Petcoke can be gasified to produce syngas. Due to their physical and chemical properties, however, the lowest-value grades of petcoke may be difficult to gasify using existing processes. Therefore, an alternative technology that will easily gasify low-value petcoke in an environmentally safe way is desireable. PEPS® appears to offer such promise.

## Alternate Gasification Technology – PEPS®

EnerSol Technologies, Inc. (EnerSol) has developed the PEPS® Gasification system, which offers certain benefits in the gasification of problematic feedstocks. The PEPS® utilizes plasma energy to enhance the gasification process. The Department of Defense (DOD), with support from the Department of Energy (DOE), has funded engineering and testing of the PEPS® Gasification system as a technology transfer opportunity to increase the domestic source of environmentally friendly transportation fuels.

In the PEPS® Gasification process, a plasma torch is used to accelerate and enhance gasification reactions, and to provide auxiliary heat to the gasifier. By

# Production from Solid and Liquid Carbonaceous Feedstocks

CTL diesel is likely to be produced with iron catalyst to take advantage of its intrinsic water-gas-shift activity and higher tolerance for trace impurities, as compared to the cobalt catalyst used in GTL diesel production. Both attributes are considered beneficial to processing coal-derived syngas which is hydrogen deficient and has more trace impurities. This difference in catalyst choice will result in slightly different chemical compositions between CTL and GTL diesel. Yet testing results show the two diesel fuels have exhibited similar emission reduction benefits in engine combustion testing (11, 12).

**Table II. Engine Emission Performance Data (Cummind ISM Engine).**

| Fuel | NOx | THC* | PM | BSFC* |
|---|---|---|---|---|
| | g/hp.h | g/hp.h | g/hp.h | g/hp.h |
| Base | 2.08 | 0.33 | 0.125 | 196.5 |
| F-T | 1.96 | 0.22 | 0.102 | 191.8 |
| % Reduction | 5.6 | 33.4 | 18.7 | 2.4** |

*THC: total hydrocarbon; BSFC: brake specific fuel consumption
**Tha lower BSFC number for F-T diesel should partially offset the disadvantage of its lower heating value.

## Conclusions

Prospects of higher crude price that began in 2004 are expected to rekindle interests in search for alternatives to supplement conventional crude oil. CTL fuels could be one of the viable options in the U.S. because of the abundant domestic coal reserves. Results from the two EECP feasibility studies at WMPI and Texaco show that both projects could be financially feasible under the project specific circumstances. Successful implementation of EECP projects will prove out the commercial viability of advanced CTL technologies operated in integrated mode. The CTL technologies will be further improved through two new contracts at ICRC/Syntroleum and HTI. CTL and GTL diesel fuels have similar attributes and have been shown to exhibit similar emission reduction benefits in engine combustion testing.

## References

1. Caruso, G.F., "Welcome and opening remarks", paper presented at the 13th Annual EIA Midterm Outlook and Modeling Conference, 12 April, 2005, Washington, DC.

2. E. Schmetz, C.L. Miller, J. Winslow, and D. Gray, Should there be a role for clean liquid transportation fuels from domestic coal in the nation's energy future?", Paper presented at the Pittsburgh Coal Conference, 12-15 September, 2005, Pittsburgh, PA. www.engr.pitt.edu/pcc/
3. Quarterly technical progress report (July to September 2004) submitted by WMPI PTY, LLC. under DOE Cooperative Agreement DE-FC26-00NT40693, November 2004
4. Early entrance coproduction plant, Phase I preliminary concept report, submitted by Texaco Energy Systems Inc. under DOE Cooperative Agreement DE-FC26-99FT40658, May 2001
5. Early entrance coproduction plant, Phase II final report, submitted by Texaco Energy Systems LLC under DOE Cooperative Agreement DE-FC26-99FT40658, January 2004
6. "Hydrogen from coal program: research, development, and demonstration plan.", issued by DOE, 26 September, 2005, www.fe.doe.gov
7. Syntroleum press release: "Syntroleum initiating testing program with coal-derived synthesis gas", 29 November, 2005.
8. "Rentech and Headwaters dissolve FT Solutions joint venture", Gasification News, February 2006, P. 5
9. Maly, R.R., "Effects of GTL diesel fuels on emissions and engine performances", Paper presented at the 2004 diesel engine emission reduction conference, 29 August – 2 September, 2004, Coronado, California
10. May, M., "Development and demonstration of Fischer-Tropsch fueled heavy-duty vehicles with control technologies for reduced diesel exhaust emission", Paper presented at the 2003 diesel engine emission reduction conference, 24-28 August, 2003, Newport, Rhode Island
11. Montalvo, D.A., "Engine evaluation of Fischer-Tropsch diesel fuel, - Phase I", report submitted to Bechtel Group, Inc. by Southwest Research Institute (SwRI-03-7502), October 1995
12. Schaberg, P.W., et al., "Diesel Exhaust Emissions Using Sasol Slurry Phase Distillate Process Fuels", SAE Paper 972898 presented at the International Fall Fuels & Lubricants meeting & Exposition, , 13-16 October, 1997, Tulsa, Oklahoma.

by President Bush in 2002 (6). One element of the program is "alternative production pathway" including the use of CTL fuels as feedstocks for hydrogen production near the point of use. In 2005, two cooperative agreements were awarded for activities in this area. The two contractors are ICRC (Integrated Concepts and Research Corporation) and HTI (Headwaters Technology Innovation). Both will work on improved production of CTL fuels with high-hydrogen content and the use of these fuels for hydrogen production and other applications. Highlights of these two projects are summarized below.

## ICRC (Integrated Concepts and Research Corporation) Project

Under the contract, ICRC and its subcontractor Syntroleum Corporation will study the CTL fuels production using F-T cobalt catalysts developed by Syntroleum. A mobile, two-staged bench-scale CSTR (continuous stirred tank reactor) unit will be assembled to perform the F-T testing with a slip syngas stream from an existing coal gasifier. The syngas feed used in these tests will have a hydrogen to carbon monoxide ratio around 2.0. A testing duration of 2500 hours is planned to obtain the catalyst performance data needed for the task of commercial CTL plant feasibility study under this contract. An important part of this work is to search for a guard bed placed before the F-T reactor to remove the trace impurities present in the cleaned syngas feed. Research quantities of finished F-T fuels will be produced and tested for diesel, jet fuel, and reforming to hydrogen applications.

Two of the host sites for F-T catalyst testing under consideration are (a) Eastman Chemical Company's chemicals from coal plant in Kingsport, Tennessee and (b) Dakota Gasification Company SNG (substitute natural gas) plant in Beulah, North Dakota. The former uses a Texaco coal gasifier and the latter Lurgi fixed-bed coal gasifiers. Both plants use Rectisol technology for syngas cleanup. More information on this project can be found elsewhere (7).

## HTI (Headwaters Technology Innovation) Project

HTI is the prime contractor of this project. Other team members include GTI (Gas Technology Institute), Nexant, Inc., UOP, Pall Corp., ANL (Argonne National Lab), and Air Force Testing Lab at Wright-Patterson Air Force Base. Under this project, a F-T PDU (process development unit) reactor system (16-inch diameter) will be designed, constructed, and operated in integrated mode with the Flex Fuel Gasifier (12 tons of coal per day) at GTI. It will have a nominal capacity of 10 barrels per day. A GTI proprietary technology using Morphysorb® solvent will be employed to clean up the syngas.

For the F-T synthesis, two catalysts will be used. The first one will be a precipitated iron catalyst supplied by a commercial vender. It will be a high-alpha catalyst to be operated in a slurry-phase reactor. The second one will be a HTI-proprietary skeletal iron catalyst. It will be a medium-alpha catalyst to be operated in an ebullated bed reactor.

UOP will perform the primary F-T products upgrading. ANL will perform the reforming of CTL fuels for hydrogen production. Air Force Lab will perform the combustion testing of F-T jet fuels.

Negotiation has been underway between HTI and DOE to add a new task for F-T PDU design/construction under this contract. This was necessitated after the withdrawal of Rentech from the contract (8).

## CTL versus GTL Diesel Fuels

GTL diesel production worldwide is projected to reach above 610,000 barrels per day by 2011 (9). It has attracted a lot of interest from both engine and fuel companies in the U.S. and overseas because of its excellent attributes. Results from the extensive engine testing have shown that use of GTL diesel, either as neat fuel or in a blend with conventional diesel, can lead to significant emission reductions over the conventional diesel (9, 10). Tables I and II show the properties of the testing fuels and their emission performance, respectively (10). These results show that GTL diesel has negligible aromatics, higher cetane number, and lower volumetric heating value when compared to the base fuel (a ultra-clean conventional diesel). It yields lower emissions than base fuel in all aspects: NOx by 5.5%, THC (total hydrocarbon) by 33.4%, and particulate matters (PM) by 18.7%. These reductions are attributed to the cleaner properties of GTL diesel. GTL diesel also has higher fuel utilization efficiency, as indicated by the lower BSFC (brake specific fuel consumption) number in Table II. This should help partially offsetting its disadvantage in lower heating value.

**Table I. GTL Diesel and Base Fuel Diesel Properties**

| Property | F-T (GTL) | Base Fuel |
|---|---|---|
| Aromatics, % | <0.1 | 8.1 |
| Density, $g/cm^3$ | 0.7836 | 0.832 |
| Cetane Number | >76 | 42.1 |
| High Heating Value, Btu/gal | 138,669* | 137,760 |
| Sulfur, ppm | <1 | 2 |

*F-T diesel has ~4% lower volumetric fuel efficiency compared to base fuel.

allowing the adjustment of reactor temperature without significantly disturbing reactor stoichiometry, the plasma torch provides an additional degree of freedom unavailable in conventional gasifiers. This novel system enables the conversion of low value, difficult grades of petcoke to produce syngas. After this syngas is properly conditioned, it can be further processed using conventional F-T technology to produce clean fuels for DOD, industrial, and consumer applications. In addition, the PEPS® can be used to produce hydrogen gas for fuel cell or transportation applications.

## History of Plasma Energy Technology

Plasma energy technology has been studied since the 1960's. Plasma torches similar to those utilized in the PEPS® system were originally developed for the aerospace industry to simulate the reentry conditions of space vehicles. The technology was also used by the specialty metals industry to produce high purity metals. During the 1980's, the Government funded considerable research to apply this technology for treating the most difficult-to-treat wastes produced by our society. There is high interest in plasma energy technology because of its ability to achieve temperatures of over 10,000°F. Such high temperatures are sufficient to melt inorganic materials while thoroughly destroying all organic compounds.

As a result of this effort, industrial plasma torches have been developed which can be utilized for high temperature processing of a variety of materials. EnerSol has successfully applied plasma torches in multiple reactor systems for the gasification of materials.

## Brief Review of PEPS® Technology and Prior Work

PEPS® technology has been applied to a number of difficult-to-manage waste streams. The DOD has sponsored the construction and demonstration of PEPS® systems for the destruction of solid, liquid, and sludge/slurry wastes. The PEPS® has successfully destroyed these wastes, with environmental performance better than EPA standards.

A typical PEPS® waste destruction system is shown below in Figure 1. Waste materials, which may be solid, liquid, or sludge/slurry, are introduced to the reactor though an appropriate feed mechanism. A controlled amount of an oxidant is co-fed to the reactor. The reactor, which is operated at near-atmospheric pressure, is heated to temperatures between 2,200°F and 3,000°F by a plasma torch. Under these processing conditions, organic materials decompose and gasify, forming primarily hydrogen ($H_2$) and carbon monoxide (CO).

Inorganic materials are melted, forming a pool inside the reactor. By controlling the composition of this molten pool, EnerSol can ensure the formation of a non-leaching glassy product (slag). Glass forming materials, such as silica and lime, may be added to the system to control the slag composition. The slag may be continuously or periodically removed from the reactor, a process referred to as tapping. When the waste feed material contains significant levels of metals, the molten pool may be allowed to stratify, permitting the separate recovery of metals and glassy slag.

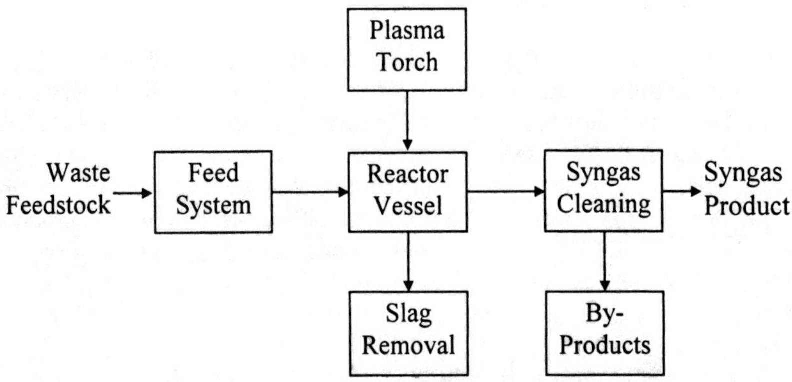


*Figure 1. Flow Diagram of a PEPS® Waste Destruction System*

Through DOD funded programs, a Mobile PEPS® waste destruction system was built and tested with a variety of feedstocks. During a recent (2003-2005) U.S. Army funded program, EnerSol upgraded and operated the Mobile PEPS®, successfully completing extensive testing on additional feed streams. The National Defense Center for Environmental Excellence (NDCEE) provided monitoring and coordination support. Three surrogate waste streams were processed, and tests were performed to measure both performance and environmental benefits. The results of the high-temperature waste destruction and gasification process were validated by the NDCEE, independent third-party sampling personnel, and outside laboratories.

Building on the success of the preceeding waste destruction PEPS® programs, the DOD tasked the NDCEE/EnerSol team to design a PEPS® Gasification system that could demonstrate the conversion of petcoke to syngas, with the focus on producing ultra-clean liquid fuels from the low-value grades of petcoke.

# Overview of PEPS® Petcoke Gasification Program

The specific goal of the Clean Fuels PEPS® project is to design a system that, when installed and operational, will demonstrate that the PEPS® Gasification process is efficient, environmentally compliant, and cost competitive for alternative fuel production with specific emphasis on using petcoke as a feedstock. The initial phase of the project was divided into three subtasks:

- Preliminary Engineering Design
- Risk Reduction Testing
- Detailed Engineering Design

The first two subtasks have been completed, and detailed engineering is in progress. A second phase of the Clean Fuels PEPS® program is expected to provide for construction and testing of the system at a F-T fuel production site.

## Preliminary Engineering Subtask

During the Preliminary Design subtask, EnerSol defined the requirements for the Clean Fuels PEPS® gasification system. Feedstock characteristics, syngas quality requirements, system sizing, and a preliminary system design were developed and documented. In addition, a few areas of technical challenge were identified, which led to the second subtask (Risk Reduction Testing). At this phase of the project, a site had not been selected, so the preliminary design was site-independent.

### *Petcoke Characteristics*

Petroleum coke is a by-product from thermal cracking (coking) operations used at approximately 57 oil refineries in the U.S (*5*). The types of petcoke that are produced vary widely depending upon the quality of the crude oil and the severity of the thermal cracking process. For the purpose of this design, EnerSol assumed that the lowest value petcoke would be fed to the PEPS® gasification system. For purposes of preliminary design, it was assumed that the petcoke would contain 5 percent sulfur and 1 percent ash (inorganics). It was further assumed that the physical form of the petcoke would be "shot coke", a particularly hard and abrasive type of petcoke. Petcoke of these characteristics has a low value, and in periods of low energy prices has in some cases, resulted in a disposal cost to the refineries.

*Syngas Specifications*

It is anticipated that Clean Fuels PEPS® unit will be constructed at an industrial site, such as an existing refinery, and that the syngas produced by the Clean Fuels PEPS® will be fed to an existing F-T process for the production of ultra-clean liquid fuels. Since the site is undefined, the specific F-T process is unknown. There are a variety of F-T processes, and the requirements for syngas vary from process to process. A primary differentiator between F-T processes is the catalyst used in the process. The catalysts used by the F-T process are primarily iron-based or cobalt-based. F-T processes using iron-based catalysts typically operate at higher temperatures and pressures, and can tolerate higher levels of impurities, than F-T processes which use cobalt-based catalysts. Some typical Fischer-Tropsch process conditions and syngas requirements are shown in Tables I and II.

**Table I. Typical Conditions for Fischer-Tropsch Synthesis**

| *Catalyst* | *Temperature* | *Pressure* | *H2:CO ratio* |
|---|---|---|---|
| Iron-based | 572 - 662°F<br>(300 - 350°C) | 145 - 580 psi<br>(10 - 40 bar) | 1.7:1 |
| Cobalt-based | 392 - 464°F<br>(200 - 240°C) | 100- 174 psi<br>(7 - 12 bar) | 2.15:1 |

**Table II. Syngas Purity Requirements for Fischer-Tropsch Processes**

| *Component* | *Typical Requirements* |
|---|---|
| Sulfur ($H_2S$, $SO_X$, $CS_2$, COS) | from < 60 ppb to < 10 ppm |
| $NO_X$ (NO, $NO_2$) | < 0.2 ppm |
| Ammonia ($NH_3$) | < 10 ppm |
| Cyanide (HCN) | < 10 ppb |
| Hydrochloric acid (HCl) | < 10 ppb |

For preliminary design purposes, EnerSol assumed that the syngas would be delivered at 500 psig, with a 2:1 H2:CO ratio. It was further assumed that the syngas would be purified to less than 0.1 ppm of sulfur compounds and to below the limits stated in Table II for HCl and nitrogen compounds. These "specifications" were for premilinary design purposes only; it was recognized that the specifications would be further defined following selection of a site, and determination of the syngas requirements for that site's F-T process.

### *Preliminary Design Capacity*

Petcoke production capacities at refineries vary significantly, ranging from 1,200 tons per day (TPD) to over 23,000 TPD (*4,5*). Gasifier systems are capital-intensive. EnerSol expects that a typical petcoke gasifier for a refinery would be sized at approximately 3,000 TPD. For the Clean Fuels PEPS® project, the intent was to design a small pilot plant to prove the technology, *not* to design a large-scale industrial unit. For preliminary design, a pilot unit capacity of 25 TPD was selected. This unit would consume ~2,100 lbs/hr of petcoke, and produce 1,700 standard cubic feet per minute (scfm) of clean syngas.

### *Process Flow System*

The Process Flow Diagram (PFD) for the Clean Fuels PEPS® is shown in Figure 2. The major process operations include: petcoke feedstock handling, reactor vessel, plasma torch, slag removal, solids removal, energy recovery, syngas purification, and pressurization of the syngas to be delivered to the F-T processing unit

The PFD includes the processing subsystems required to convert petcoke into high quality syngas for production of liquid fuels via F-T. The PFD will remain essentially unchanged over a wide range of system capacities. However, the type of equipment selected for each subsystem will vary as system capacities vary. For example, the technologies chosen for sulfur removal in the Syngas Purification subsystem may change with system capacity. For a 2,500 TPD unit, $H_2S$ may be removed from the syngas by an amine system. For a 25 TPD pilot plant system, the high capital costs associated with an amine system may not be justified, and different technologies may be selected.

Similarly, the PFD will not vary significantly based on the system installation site. However, as noted earlier, the syngas quality specifications are dependent on the installation site and the F-T process used. Consequently, the types of equipment selected for specific subsystems may change based on the system installation site.

During preliminary design, EnerSol evaluated a variety of technologies for each of the subsystems required for the 25 TPD pilot unit. Design criteria for the subsystems in Figure 2 are briefly described below.

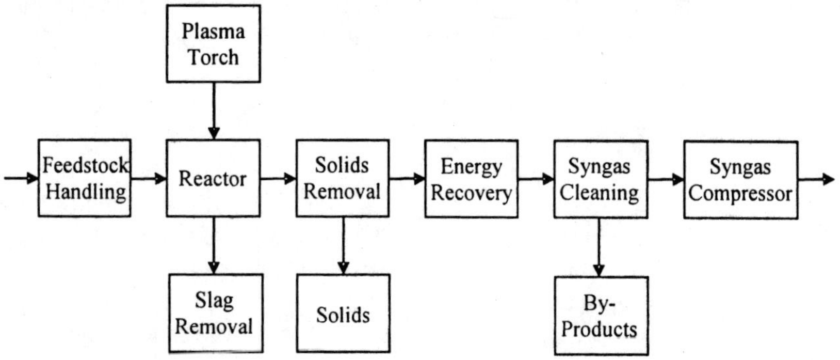


*Figure 2. Flow Diagram of PEPS® 25 TPD Gasification System*

*Process Description of 25 TPD Pilot Unit*

The petcoke is expected to be received in lump form, with most lumps sized between 4 and 6 inches. The Clean Fuels PEPS® feedstock handling equipment will provide for crushing, grinding, and feeding of the petcoke. The petcoke will be ground to a fine powder before it is fed to the reactor. The low-value grades of petcoke are quite hard and abrasive; proper design of the grinding equipment is critical for reliable operation. The preliminary design was based on pneumatic feeding of the ground petcoke, using steam as the motive gas. Some other key issues in the feedstock handling system design include determination of the desired particle size and ensuring reliable feed of the petcoke into a pressurized reactor.

The plasma assisted reactor is designed to maximize petcoke conversion while minimizing fine particulate carryover. Several reactor configurations were evaluated, and an entrained flow, vertical down-flow reactor was selected. A sketch of this style reactor is shown in Figure 3. In this configuration, a non-transferred DC arc plasma torch is mounted at the top of the refractory-lined reactor. Petcoke and oxygen are introduced through one or more nozzles mounted near the torch. The plasma torch generates a region of high temperature, highly reactive species, which interact with the petcoke and oxygen, enhancing and accelerating reactions. As the feedstocks flow downward through the reactor, the petcoke is rapidly converted to syngas. The syngas exits through a port on the side of the reactor, near the bottom. Inorganic materials collect in a molten slag bath at the base of the reactor.

To ensure good petcoke conversion, the reactor must have sufficient volume to ensure adequate gas phase residence time for petcoke conversion to syngas. A key design issue was determining the correct reactor volume.

Another significant issue was determining an appropriate reactor pressure for the 25 TPD pilot unit. As discussed earlier, the syngas product is to be delivered at 500 psig. Ideally, if the reactor were operated at a pressure in excess of 500 psig, there would be no need for the syngas compressor shown in Figure 2. Significant cost savings could be achieved by eliminating the compressor. Conversely, operation of the reactor at atmospheric (or sub-atmospheric) pressures would require the use of large, expensive, energy-intensive compressors to raise the product syngas pressure to 500 psig.

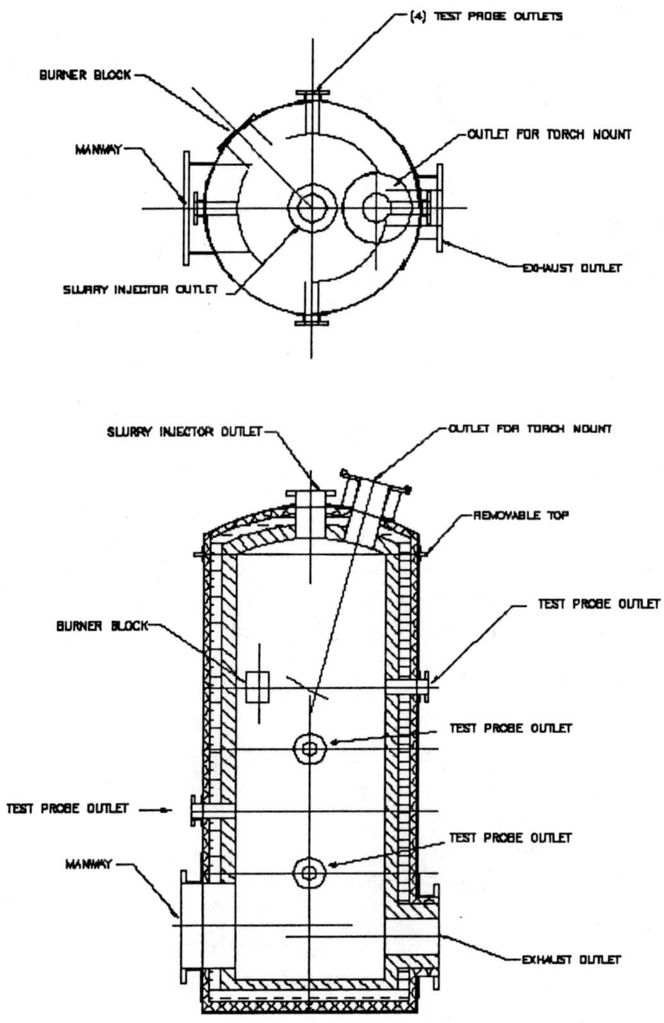


*Figure 3. Preliminary Reactor Design for PEPS® 25 TPD Gasification System*

The limiting factor in the selection of a system pressure was the plasma torch. Industrial experience with the operation of non-transferred DC plasma torches at elevated pressures is quite limited. For the 25 TPD Clean Fuels PEPS® pilot unit, a moderate pressure of 50 psig was selected as a design basis.

Although the reactor is designed to maximize petcoke conversion to syngas, there will be some solids in the product syngas. These solids will include unreacted petcoke particles and entrained ash (inorganic constituents). A cyclone was selected for solids removal downsteam of the reactor.

Recovery of a portion of the heat energy available in the hot (>2000°F) syngas stream of the 25 TPD unit is intended to improve the net thermal efficiency of the process by installing a waste heat boiler, which will produce steam for use in process heating. In a larger gasification system, additional energy recovery equipment could be installed to further improve efficiency. However, the capital costs of such equipment are not justified for a small 25 TPD pilot unit.

The syngas cleaning system is designed to condition the syngas produced in the PEPS® reactor to meet the requirements of the F-T process. The most significant functions include removal of sulfur compounds (primarily hydrogen sulfide, $H_2S$), adjustment of $H_2$:CO ratio, and removal of trace impurities. As noted above, although an amine unit might be selected for $H_2S$ removal in a larger system, this technology was deemed too expensive for the 25 TPD pilot unit. A wet scrubber was chosen for primary $H_2S$ removal for the pilot unit. A water gas shift reactor was designed for adjustment of $H_2$:CO ratio. For this pilot unit, scavenger beds were selected for removal of residual trace impurities. Syngas exiting the cleaning system will meet all purity requirements for use in a F-T process.

Although the Clean Fuels PEPS® design is intended to produce syngas for F-T synthesis of liquid fuels, it should be noted that the system, with minor modifications, could be used to produce hydrogen as the primary product. The water gas shift reactor could be modified to drive the gas composition to primarily $H_2$ and $CO_2$. The addition of a hydrogen separation unit downstream of the shift reactor would permit recovery of hydrogen as an alternate product stream.

To meet the pressure requirement of the F-T process, the syngas must be pressurized to 500 psig. Since compressor power requirements are related to the compression ratio (pressure out / pressure in), the power required can be reduced by increasing the suction pressure. As discussed earlier, concerns about plasma torch operations under 500 psig pressure resulted in a design pressure of 50 psig for the reactor.

During the Preliminary Design, a few areas of technical challenge (risk) were identified, which led to the following subtask (Risk Reduction Testing).

**Risk Reduction Testing Subtask**

Despite considerable research and careful design efforts, the design of a prototype pilot plant always involves some unknown parameters. As noted in the preliminary design discussions above, a few areas of technical challenge were identified. To build the pilot plant without investigating these issues would increase the risk. The pilot unit could experience problems and have difficulty achieving the goal of successfully demonstrating the PEPS® gasification system if these issues are not addressed. To address these challenges, four risk areas were identified. The four areas of concern were:

- Reactor design issues
- Feedstock handing issues
- Pressurized plasma torch operation
- Petcoke grinding issues

Risk Reduction Testing (RRT) was performed to mitigate risk in the first two risk areas. A temporary entrained flow, vertical down-flow reactor, dimensionally similar to the planned 25 TPD reactor, was fabricated and connected to the existing Mobile PEPS® gas cleaning system. A plasma torch and instrumentation were mounted on the temporary reactor. Petcoke was gasified in the reactor while operating over a wide range of feedrates and temperatures, and data was collected. Due to limitations of the Mobile PEPS® gas cleaning system, the reactor could not be pressurized. All RRT operations were conducted at near atmospheric pressure.

Below, each risk area will be discussed. Where available, results from Risk Reduction Testing will be presented. (Risk Reduction Testing has not been completed for the last two risk areas.)

*Reactor Design Issues*

An entrained flow, vertical down-flow reactor design was selected for the 25 TPD pilot unit. A key parameter for the design of this unit was the reactor volume required (i.e., gas residence time). Operations were conducted at a variety of conditions, and the reactor product gas was sampled to determine the mass flow of unreacted petcoke. Mass balances were performed to calculate the petcoke conversion rate.

Figure 4 presents the results of one such study. Petcoke carbon conversion was plotted versus reactor gas-phase residence time calculated for the trial runs, and a trend line was extrapolated to 100% conversion. From this graph, it was

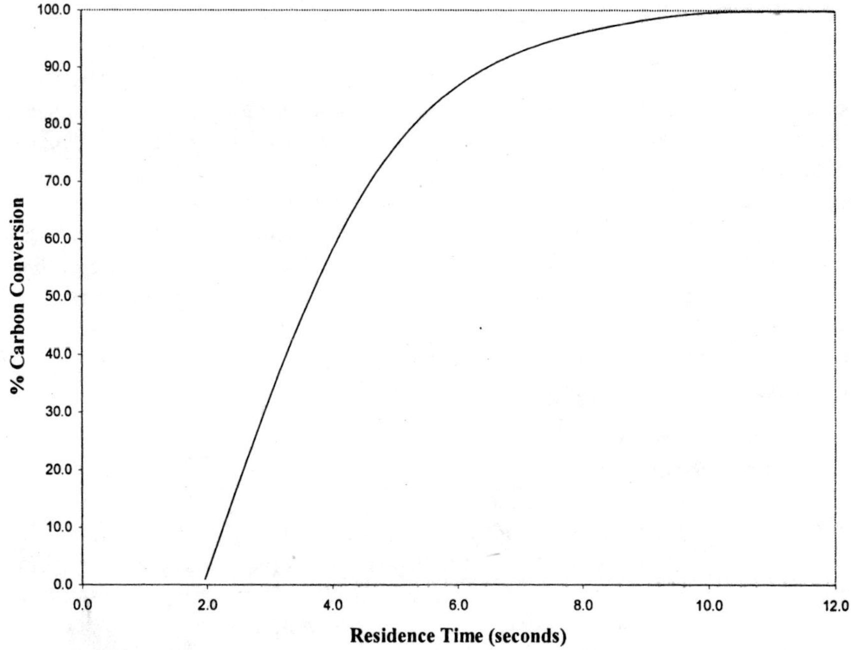


*Figure 4. RRT Study of Reactor Residence Time vs. Conversion*

concluded that approximately 10 seconds of residence time would be required for carbon conversion to approach 100% at the reactor operating conditions tested.

Additional reactor design issues studied during RRT included the effect of petcoke particle size on conversion,and the effect of reactor temperature on gas composition and $H_2$:CO ratio.

*Feedstock Handling Issues*

During the preliminary design effort, EnerSol selected pneumatic conveyance as the feed method. After further investigation, feeding the petcoke in a petcoke:water slurry was determined to be the preferred method. Although slurry feeding offered several advantages over pneumatic feeding, the introduction of significant amounts of liquid water into the reactor adversely impacted the overall thermal efficiency of the process. (Significant amounts of heat would be required to boil the excess water.) RRT efforts focused on determining the optimal (maximum) petcoke:water ratio, and investigating issues surrounding erosion and plugging in the slurry feed system.

Petcoke:water slurries were prepared at mass ratios ranging from 30:70 to 70:30. EnerSol determined that mixtures of up to 50:50 were pumpable and fed without problems to the RRT reactor. Mixtures with higher petcoke concentrations were difficult to maintain in suspension. Experimentation with surfactants and stabilizers resulted in a pumpable mixture at 65% petcoke and 35% water.

Injection of the petcoke:water slurry presented an additional challenge. An EnerSol-proprietary atomizing feed nozzle was designed and fabricated. It delivered excellent slurry atomization and reliability, while withstanding the high reactor temperatures and the abrasive slurry.

### *Pressurized Plasma Torch Operation*

As discussed earlier, there is limited experience with the operation of industrial plasma torches in pressurized reactors. Testing of a torch under 50 psig (and higher) pressure prior to construction of the 25 TPD pilot unit would be advantageous.

### *Petcoke Grinding Issues*

Petcoke can be quite hard and abrasive. In order to design and specify a grinder for this service, some bench-scale test grinding is necessary. EnerSol expects that this testing will be completed prior to the final design and procurement of a petcoke grinder.

## Future Work on Clean Fuels PEPS®

Detailed design of the Clean Fuels PEPS® project is underway. Additional Risk Reduction Testing has been defined. A second phase of the Clean Fuels PEPS® project is expected to provide funding for construction and testing of the system.

## Summary

The PEPS® Gasification system offers the capability to convert a petroleum refinery by-product into valuable, ultra-clean liquid fuels. Alternately, the system can be modified to generate hydrogen, another valuable clean fuel. These fuels could reduce the nation's dependence on imported oil. Preliminary design for a 25 TPD Clean Fuels PEPS® pilot unit has been completed, as has some risk reduction testing. Future funding for completion of the project is anticipated.

PEPS® technology has previously been proven to be safe and effective for the treatment of wastes and by-products.

## Acknowledgements

This project is being funded and conducted by the NDCEE, which is managed by the Office of the Assistant Secretary of the Army (Installations and Environment) and operated by Concurrent Technologies Corporation. EnerSol Technologies, Inc. is providing system design. The Department of Energy is providing Technical Monitoring.

## Disclaimer

The views, opinions, and/or findings contained in this paper are those of the authors and should not be construed as an official Department of Army position, policy, or decision unless so designated by other official documentation.

## References

1. Energy Information Administration, Annual Energy Outlook 2005, January 2005, http://www.eia.doe.gov/oiaf/archive/aeo05/pdf/0383(2005).pdf
2. Energy Information Administration, US Natural Gas Prices, April 2006 http://tonto.eia.doe.gov/dnav/ng/ng_pri_sum_dcu_nus_a.htm
3. Energy Information Administration, Supply and Disposition, US Petroleum Coke, April 2006 http://tonto.eia.doe.gov/dnav/pet/pet_sum_snd_c_nus_eppc_mbbl_a.htm
4. Energy Information Administration, Definitions, Sources and Explanatory Notes, April 2006 http://tonto.eia.doe.gov/dnav/pet/TblDefs/pet_sum_snd_tbldef2.asp
5. Energy Information Administration, Petroleum Supply Annual 2004, Volume 1, Released June 2005, Table 38, http://www.eia.doe.gov/pub/oil_gas/petroleum/data_publications/petroleum_supply_annual/psa_volume1/current/pdf/table_38.pdf

# Chapter 4

# Effect of Ultrasonic Irradiation on Enxymatic Transesterification of Waste Oil to Biodiesel

**Hong Wu and Min-hua Zong**

**Lab of Applied Biocatalysis, South China University of Technology, 1 Wushan Street, Guangzhou 510640, People's Republic of China**

The effect of ultrasonic irradiation on Novozym435-catalyzed transesterification of waste oil to biodiesel in a solvent free system was explored. The application of ultrasonic irradiation instead of shaking during the transerterification led to no significant improvement in the reaction rate. However, Novozym 435-catalyzed transesterification of waste oil was markedly accelerated by pretreatment of the enzyme with ultrasonic irradiation at 20 kHz, 80 W for 30 min. During the three-step batch reaction with pretreated Novozym 435, the reaction time of each step reduced to 5 h, 5 h and 10 h from original 6 h, 8 h and 16 h. On the other hand, Novozym 435 pretreated with ultrasonic irradiation showed good operational stability as indicated by a slight loss in activity after 10-batch operation.

Biodiesel (long-chain monoalkyl fatty acid esters), a renewable and green fuel, is made from vegetable oil or animal fats. At present, the high cost of biodiesel, of which the raw material amounts to 75%, prohibits its wide application. Compared with chemical method, enzymatic process[1,2] seems to be a promising alternative because of its mild reaction conditions, easy recovery of product, being free of chemical wastes and low demangding on raw materials, which makes it possible to use waste oil as substrate for enzymatic production of biodiesel.

Waste oils, from restaurants and household disposals and being creating serious problems of environmental control and food safety, have been considered as good raw material for biodiesel production. Immobilized *Candida antarctica* lipase was found to be effective for the methanolysis of waste oil.[2,3] A three-step methanolysis protocol could be used to protect lipase from inactivation by methanol. Compared with one-step reaction, it needs a longer time to reach the reaction equilibrium. So, efforts should be made to increase enzymatic reaction rate. Reports on the enhancement of the activity of certain enzymes by applying ultrasonic irradiation on the enzymes[4,5] led us to investigate its effects on the enzymatic transesterification of waste oil to biodiesel in a solvent free system.

## Experimental

### Materials

Waste oil was collected from the restaurant in South China University of Technology. The saponification value was 200.3 mg KOH/g, from which, the average molecular weight of the waste oil was known to be 840.2. Novozym 435 (lipase B from *Candida antarctica*, 164 U/g, 1unit corresponds to the amount of enzyme that produces 1 μmol methyl oleate from triolein per minute at 35 °C) was kindly donated by Novozymes Co. (Denmark). Methyl palmitate, methyl stearate, methyl oleate, methyl linoleate, methyl linolenate and methyl heptadecanoate (as an internal standard) were purchased from Sigma (USA). All other chemicals were also obtained commercially and of analytical grade. Ultrasonic irradiation experiments were carried out using an ultrasonic bath (Type NP-B-400-15; Newpower Co. Ltd., China).

### Determination of power density

Power density at sample sites in the ultrasound bath was determined according to the method described by Barton.[6]

### Reaction

The reaction was carried out in 50-ml flask capped with a septum, incubated at 40 °C and in the presence or absence of ultrasonic irradiation. The reaction mixture contained 66 U Novozym 435, 10 g waste oil and 0.38 g methanol (1 molar equivalent). Samples of 100 μl were periodically withdrawn and centrifuged (12000 rpm, 10 min) and the upper layer was mixed with methyl heptadecanoate for GC analysis. In the ultrasonic pretreatment experiment, Novozym 435 was immerged into 10 g waste oil and exposed to ultrasonic irradiation for a predetermined period of time.

### Analysis

The methyl ester (ME) content of the reaction mixture was assayed with a GC2010 gas chromatography (Shimadzu Corp., Kyoto) using a flame ionization detector and a HP-5 capillary column (0.53 mm × 15 m, HP, USA). The column temperature was hold at 180 °C for 1 min, raised to 186 °C at 0.8 °C /min and kept for 1 min, then upgraded to 280 °C at the rate of 20 °C /min. Nitrogen was used as the carrier gas at 2 ml/min. Split ratio was 1:100 (v/v). The injector and the detector temperatures were set at 250 °C and 280 °C, respectively.

## Results and Discussion

The use of an ultrasonic bath has the advantages of simplicity, low cost, and modest power outputs reducing the likelihood of localized heating effects with ultrasound probe and cuphorn systems. The most obvious limitations are that the operating frequency is normally fixed and power densities vary within the bath. Consequently standardization of sample locations is essential for comparative purposes. In our study, the water addition to the bath was as high as 230 mm from the bottom and the sample was fixed at the position of 180 mm from the bottom. The maximum power density with water as medium was obtained in the center and this location was used for all the following reactions. The power densities were 0.21W/ml, 0.28 W/ml, 0.36 W/ml and 0.48 W/ml, respectively, when the power outputs were 40 W, 60 W, 80 W and 100 W, respectively.

As can be seen in **Figure 1**, the reaction accelerated markedly with the increase in power output when it was below 80 W. Further increase in power output up to 100 W, however, resulted in little rise in conversion after reaction for 2 h and the ME content of the product was lower than 30% in spite of the higher initial reaction rate, suggesting partial inactivation of the enzyme. This has been proved by experiments showing that the enzyme retained only 81% of its original activity after being treated with ultrasonic irradiation at 100 W. However, the best result achieved was only comparative to the control which was carried out with shaking at 150 rpm. This is mainly due to the high viscosity of waste oil and little enhancement in mass transfer at a low ultrasonic intensity and inactivation of enzyme at a high ultrasonic intensity.

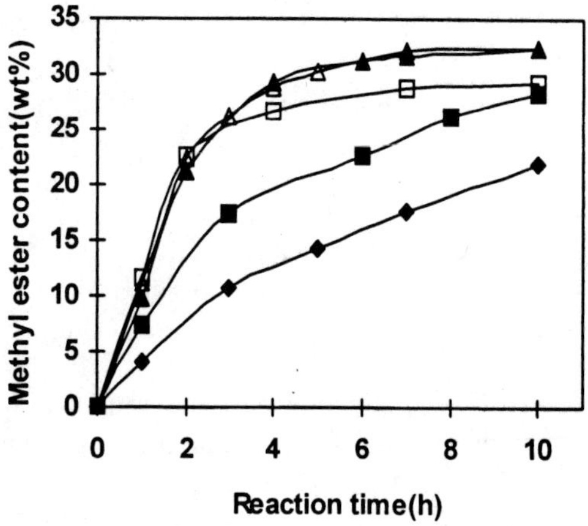


*Figure 1. Effect of ultrasonic power on enzymatic transesterification of waste oil. The reactions were carried out at 40±2 °C in the presence of ultrasonic irradiation and the reaction mixture contained 10 g waste oil, 0.38 g methanol and 66 U Novozym 435. Control was carried out at 40 °C and 150 rpm. □,40 W; ■,60 W; ▲,80 W; △,100 W; ◆, control.*

It has been reported that the reaction rate could be enhanced by pretreatment of enzyme with ultrasonic irradiation.[7] In our study, It was found that the reaction proceeded quite slowly during the first 1 hour of reaction as indicated by the low ME content of the reaction mixture (10.9%) and then accelerated. It seems that there exists a period for the enzyme to be activated. So, we speculated that pretreatment of enzyme with ultrasonic irradiation might activate the enzyme and accelerate the enzymatic reaction. The results depicted in **Table I** well supports the hypothesis.

Since the time enzyme was pretreated with ultrasonic irradiation could influence the reaction rate, effect of pretreatment time on the reaction was therefore investigated. As shown in **Table II**, the reaction rate increased with increasing pretreatment time up to 30 min. Further increase in the pretreatment time beyond 30 min, however, resulted in little change in reaction rate. So, the optimum pretreatment time was thought to be 30 min.

**Table I. Effect of ultrasonic pretreatment of the enzyme on enzymatic transesterification of waste oil**

| Pretreatment means | Methyl ester content (wt%) |
|---|---|
| Ultrasonic irradiation | 19.4 |
| Shaking | 12.1 |
| Control* | 10.9 |

Novozym 435 (66 U) was immerged into 10 g waste oil and then pretreated with ultrasonic irradiation (20 kHz, 80 W) at 40 °C for 30 min or with shaking at 40 °C and 150 rpm for 30 min. Methanol (0.38 g) was added to the reaction mixture and the reactions were carried out at 40 °C and 150 rpm for 1 h.

*Without pretreatment

**Table II. Effect of ultrasonic pretreatment time on enzymatic transesterification of waste oil**

| Time (min) | Methyl ester content (wt%) |
|---|---|
| 10 | 13.6 |
| 20 | 17.1 |
| 30 | 19.4 |
| 40 | 19.5 |
| 60 | 19.1 |

Novozym 435 (66 U) was immerged into 10 g waste oil and pretreated with ultrasonic irradiation (20 kHz, 80 W) at 40 °C for a certain period of time. Methanol (0.38 g) was added to the reaction mixture and the reactions were carried out at 40 °C and 150 rpm for 1 h.

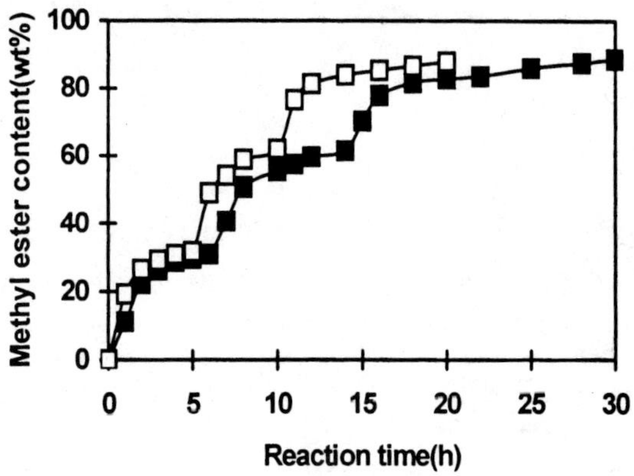


*Figure 2. Time course of pretreated Novozym 435-catalyzed transesterification of waste oil. Reaction was performed at 40 °C, 150 rpm by adding 0.38 g methanol into the mixture containing 10 g waste oil and 66 U Novozym 435 pretreated or non- pretreated by ultrasonic irradiation (20 kHz, 80 W) at 40 °C for 30 min. 0.38 g Methanol was added at 5 and 10 h or at 6 and 8 h and glycerol was rinsed with acetone after each step.* □*, pretreated;* ■*, non-pretreated.*

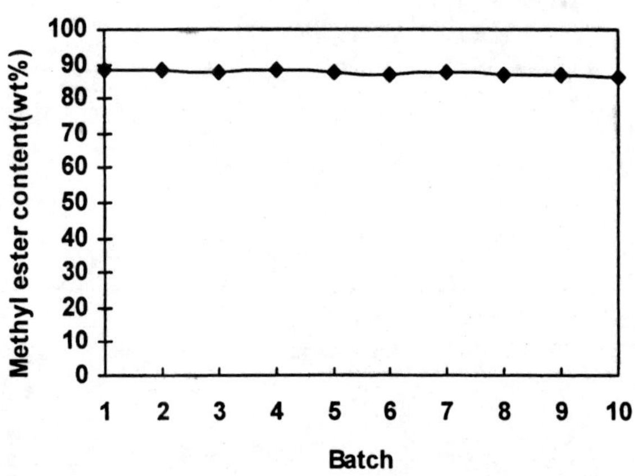


*Figure 3. Operational stability of pretreated Novozym 435. Reaction conditions were the same as in Figure 2 and three steps served as one batch with a total reaction time of 20 h.*

**Figure 2** depicts the time course of pretreated Novozym 435-catalyzed transesterification of waste oil. During the three-step batch reaction, the reaction time of each step reduced to 5 h, 5 h and 10 h from original 6 h, 8 h and 16 h and one third of the total reaction time was saved. On the other hand, Novozym 435 pretreated by ultrasound showed good operational stability as indicated by a slight loss in activity after 10-batch operation (**Figure 3**).

## Conclusion

Novozym 435-catalyzed transesterification of waste oil was markedly accelerated by pretreatment of the enzyme with ultrasonic irradiation at 20 kHz, 80 W for 30 min and Novozym 435 pretreated with ultrasonic irradiation showed good operational stability and so are potential for biodiesel production.

## Acknowledgment

This work was financially supported by Science and Technology Project of Guangdong Province (Grant No. 2003C33102), Science and Technology Project of Guangzhou (Grant No. 2005Z3-D0471).

## References

1. Steinke, G.; Kirchhoff, R.; Muknerjee, K. D. *J. Am. Oil Chem. Soc.*, **2000**, *77* (4), 361-366
2. Shimada, Y.; Watanabe, Y.; Sugihara, A.; Tominaga, Y. J. *Mol. Catal. B: Enzym.*, 2002, *17* (3-5), 133-142
3. Wu, H.; Zong, M. H.; Lou, W. Y. *Chin. J. Catal.*, **2004**, *25* (11), 903-908
4. Vulfson, E. N.; Sarney, D. B.; Law, B. A. *Enzyme Microb. Technol.*, **1991**, *13* (2), 123-126
5. Sakakibara, M.; Wang, R.; Takahashi, R.; Takahashi, K.; Mori, S. *Enzyme Microb. Technol.*, **1996**, *18*, 444-448
6. Barton, S.; Bullock, C.; Weir, D. *Enzyme Microb. Technol.*, **1996**, *18*, 190-194
7. Zong, M. H.; Du, W.; Li, H. Q.; Liu, Y. *J South China Univ. Tech.*, **2000**, *28* (3), 101-104

## Chapter 5

# Improving Enzymatic Transformation of Waste Edible Oil to Biodiesel by Adding Organic Base

**Zhi-feng Chen, Min-hua Zong*, and Hong Wu**

**Lab of Applied Biocatalysis, South China University of Technology, 1 Wushan Street, Guangzhou 510640, People's Republic of China**

The transesterification of waste edible oil and methyl acetate to biodiesel catalyzed by the immobilized lipase Novozym 435 was explored. Novozym 435 could catalyze the transesterification of waste oil with high free fatty acid (FFA) content (up to 27.8%) and methyl acetate, and methyl ester (ME) yield reached 77.5% after a reaction time of 24 h, which was much lower than that with refined corn oil being the raw material (86.2%). FFA was demonstrated to be the major influential factor on the reaction, ME yield dropped sharply with increasing FFA concentration. Acetic acid, the by-product formed in the transesterification of FFA with methyl acetate, was found to be responsible for the decrease of ME yield. Addition of organic base trihydroxymethyl aminomethane and triethylamine at the concentration of 5% based on oil weight to the reaction system could not only speed up the reaction, but improve ME yield.

## Introduction

Wastes edible oil, originated from restaurants and household disposals, is creating serious problems of environmental control and food safety. Production of biodiesel with waste edible oil as feedstock not only could reduce disposal problems, but, more importantly, would decrease the cost of biodiesel.

Presently industrial production of biodiesel from waste edible oil is performed by chemical alkaline catalysts, but it has been found that high content of free fatty acid (FFA) in waste edible oil (FFA>2%) would decrease the yield sharply due to soaps formed in the process (*1*). Several studies showed that enzymatic methanolysis with waste edible oil was a promising alternative owing to its mild reaction condition and free of chemical waste (*2*, *3*). However, this conventional protocol was associated with many drawbacks such as deactivation of lipase caused by methanol and absorption of glycerol to lipase surface, thus resulting in serious negative effect on the reaction (*4*, *5*).

Du *et al.* (*6*) have recently reported that methyl acetate was an effective acyl acceptor for biodiesel production. To the best of our knowledge, the biodiesel production from waste edible oil with methyl acetate as the acyl acceptor has not yet been reported. Therefore, the transesterification of different kinds of waste edible oil to biodiesel with methyl acetate in a solvent free system was explored in this paper and the major influential factor on the reaction was further investigated.

## Experimental

### Materials

One waste edible oil (WDO-1) was collected from fast food restaurant, the other (WDO-2) was obtained from local restaurant waste oil pool and pretreated with activated clay. The refined corn oil (RCO) was purchased from local supermarket. Novozym 435 (lipase B from *Candida antarctica*, 164 U/g, 1unit corresponds to the amount of enzyme that produces 1 μmol methyl oleate from triolein per minute at 35 °C) was kindly donated by Novo Nordisk Co. (Denmark). Methyl palmitate, methyl stearate, methyl oleate, methyl linoleate, methyl linolenate and methyl heptadecanoate (as an internal standard) were purchased from Sigma (USA). All other chemicals were also obtained commercially and of analytical grade.

### Reaction

The reaction was carried out at 40 °C and 150 rpm. The reaction mixture contained 5 g waste edible oil, 6.93 g methyl acetate with the molar ratio of methyl acetate to total fatty acid being 16:3 and 246 U Novozym 435. The reaction conditions of the three-step methanolysis were the same as our previous study (7). Samples of 5 μl were periodically withdrawn and mixed with methyl heptadecanoate for GC analysis.

### Analysis

The methyl ester (ME) yield in the reaction mixture was assayed with a GC2010 gas chromatography (Shimadzu Corp., Kyoto) equipped with a HP-5 capillary column (0.53 mm × 15 m, HP, USA). The column temperature was hold at 180 °C for 1 min, raised to 186 °C at 0.8 °C /min and kept for 1 min, then upgraded to 280 °C at the rate of 20 °C /min. Nitrogen was used as the carrier gas at 2 ml/min. Split ratio was 1:100 (v/v). The injector and the detector temperatures were set at 250 °C and 280 °C, respectively. The water content of the waste edible oil and refined corn oil were measured by Karl Fischer titration using 787 KF Titrino (Metrohm Ltd. Switerland). The FFA content and the Saponification value were determined by GB/T 5530-1998 and GB 5534-85, respectively.

## Results and Discussion

The transesterification of different oil with methyl acetate was first explored (**Figure 1**). The ME yield of WDO-1 was close to that of RCO after reaction for 24 h. However, the ME yield of WDO-2 was only 77.5%, which was lower than that of WDO-1 (85.2%) and RCO (86.2%).

To determine the reason for the differences in ME yield, some important parameters of three kinds of oil were measured and depicted in **Table I**. The major difference was found to be in the content of FFA. So, the effect of the FFA on the reaction was subsequently investigated by adding different amounts of oleic acid into the refined corn oil.

As depicted in the **Figure 2**, the higher the FFA content, the lower the ME yield, which was similar to the result achieved in **Figure 1**, suggesting that the FFA content was the key influential factor on the reaction. However, the initial reaction rate was not significantly different, indicating that FFA showed no negative effect to the enzyme activity. Thus, acetic acid, the by-product of the process, formed in the transesterification of FFA with methyl acetate might be responsible for the decrease of the ME yield. Different amounts of acetic acid were then added directly into the RCO and the results obtained well supports the hypothesis (shown in **Figure 3**). The reaction rate and the ME yield decreased with increasing acetic acid content. This can be explained by the change of pH in the micro-environment of enzyme and thus resulted in the inhibition of enzyme activity.

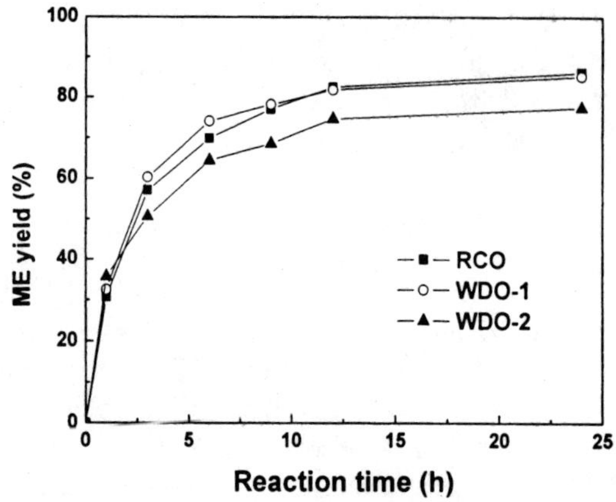


*Figure 1. Time course of the reaction. The reactions were carried out at 40 °C and 150 rpm and the reaction mixture contained 5 g waste edible oil, 6.93 g methyl acetate and 246 U Novozym 435.*

**Table I. Main parameters of waste edible oil and refined oil**

| | WDO-1 | WDO-2 | RCO |
|---|---|---|---|
| Water content (ppm) | 320 | 300 | 250 |
| FFA content (%) | 4.6 | 27.8 | 0.1 |
| Saponification value (mg KOH/g of sample) | 193 | 191 | 187 |

To decrease the negative effect of acetic acid, some efforts have been tried by addition of different organic bases, such as trihydroxymethyl aminomethane (TB), triethylamine (TEA) and pyridine (PD) into the reaction mixture of WDO-2 at the beginning of the transesterification As shown in **Figure 4**, the reaction rate was markedly enhanced during the process, on the other hand, the improvement of the ME yield was also observed. Among the organic bases studied, TB was found to be the best for the reaction.

A comparative study of biodiesel production with WDO-2 using three-step methanolysis and one-step transesterification with methyl acetate was also conducted here. As can be seen in **Figure 5**, the ME yield of the three-step methanolysis after reaction for 72 h was 69.1%, which was much lower than those of the one-step transesterification with or without addition of organic base depicted in **Figure 4**, suggesting that the enzymatic transesterification with methyl acetate was more effective than the enzymatic methanolysis in solvent free system for biodiesel production.

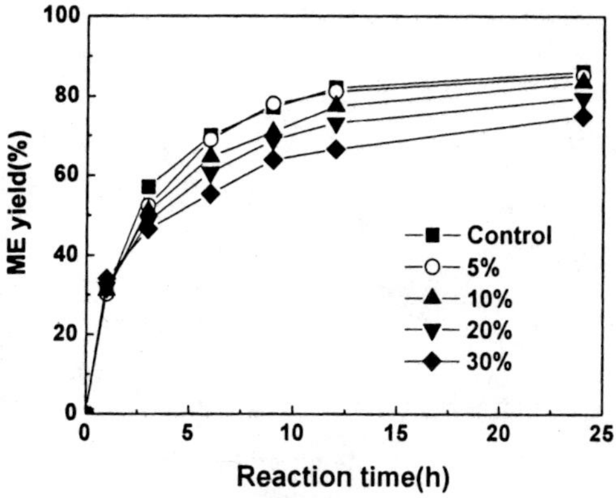


*Figure 2. Effect of free fatty acid on the reaction. The reactions were carried out at 40 °C and 150 rpm and the reaction mixture contained 5 g mixture of RCO and oleic acid, 6.93 g methyl acetate and 246 U Novozym 435.*

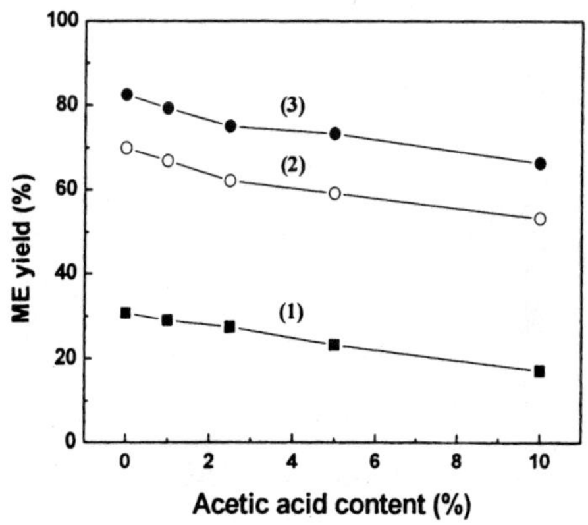


*Figure 3. Effect of acetic acid on the reaction. The reactions were carried out at 40 °C and 150 rpm and the reaction mixture contained 5 g RCO, 6.93 g methyl acetate and 246 U Novozym 435. Reaction time: (1) 1 h, (2) 6 h, (3) 12 h*

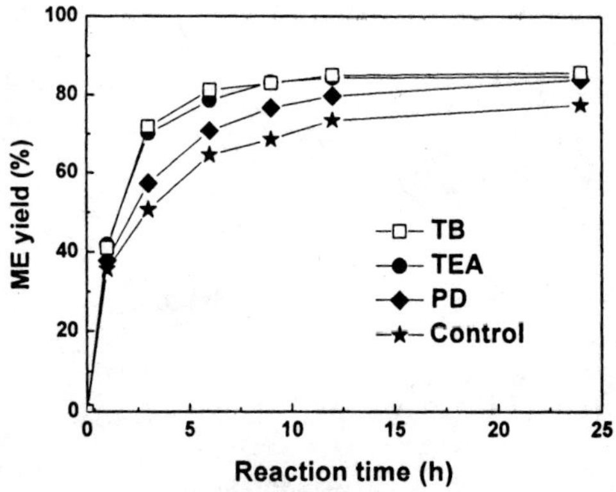


*Figure 4.* ***Effect of organic base on the reaction.*** *The reactions were carried out at 40 °C and 150 rpm and the reaction mixture contained 5 g WDO-2, 6.93 g methyl acetate, 246 U Novozym 435 and 0.25 g different organic base.*

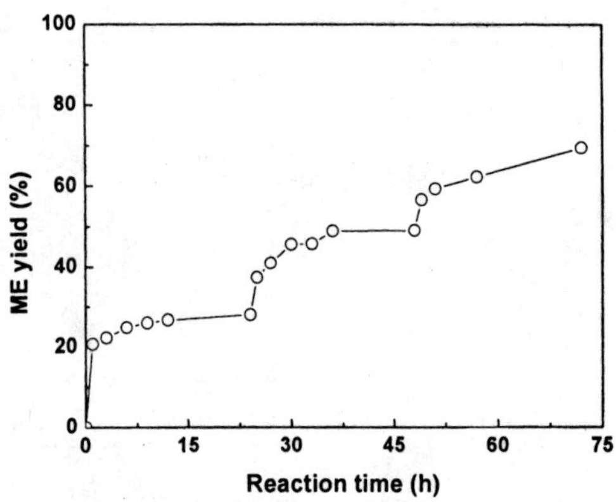


*Figure 5. Time course of three-step enzymatic methanolysis of WDO-2. Reaction was performed at 40 °C, 150 rpm by adding 0.38 g methanol into the mixture containing 5 g WDO-2 and 66 U Novozym 435, 0.38 g Methanol was added at 24 and 48 h.*

## Conclusion

Novozym 435, immobilized lipase from *Candida antarctica*, type B could efficiently catalyze the transesterification of waste edible oil with methyl acetate. However, FFA affected the reaction clearly, and the ME yield drops sharply with increasing FFA concentration. Acetic acid, the by-product formed in the transesterification of FFA with methyl acetate, was proved responsible for the decrease of reaction rate and ME yield. Addition of organic base to the reaction system could not only speed up the reaction, but also improve the ME yield. This protocol seems to be an efficient alternative for the conversion of waste edible oil to corresponding methyl esters.

## Acknowledgment

This work was financially supported by Science and Technology Project of Guangdong Province (Grant No. 2003C33102), Science and Technology Project of Guangzhou (Grant No. 2005Z3-D0471).

## References

1. Ramadhas, A. S.; Jayaraj, S.; Muraleedharan, C. *Fuel*, **2005**, *84* (4), 335-340
2. Wu, H.; Zong, M. H; Luo, Q.; Wu, H. C. *Prepr. Pap. - Am. Chem. Soc., Div. Fuel Chem.*, **2003**, *48* (2), 533-534
3. Watanabe, Y.; Shimada, Y.; Sugihara, A.; Tominaga, Y. *J. Am. Oil Chem. Soc.*, **2001**, *78* (7), 703-707
4. Shimada, Y.; Watanabe, Y.; Samukawa, T.; Sugihara, A.; Noda, H.; Fukuda, H.; Tominaga, Y. *J. Am. Oil Chem. Soc.*, **1999**, *76* (7), 789-792
5. Dossat, V.; Combes, D.; Marty, A. *Enzyme Microb. Technol.* **1999**, *25* (3-4), 194–200
6. Xu, Y. Y.; Du, W.; Liu, D. H.; Zeng, J. *Biotechnol. Lett.* **2003**, *25* (15), 1239-1241
7. Wu, H.; Zong, M. H.; Lou, W. Y. *Chin. J. Catal.*, **2004**, *25* (11), 903-908

## Chapter 6

# Performance and Characterization of NiO–MgO Solid Solution Modified with Noble Metals in Oxidative Steam Reforming of Methane under Pressurized Conditions

**Shigeru Kado[1], Mohammad Nurunnabi[1], Yuya Mukainakano[1], Tomohisa Miyazawa[1], Kenji Nakao[1], Kazu Okumura[2], Toshihiro Miyao[3], Shuichi Naito[3], Kimihito Suzuki[4], Ken-ichiro Fujimoto[4], Kimio Kunimori[1], and Keiichi Tomishige[1]**

**[1]Institute of Materials Science, University of Tsukuba, 1–1–1 Tennodai, Tsukuba-shi 305–8573, Ibaraki, Japan**
**[2]Department of Materials Science, Faulty of Engineering, Tottori University, Tottori, Japan**
**[3]Department of Applied Chemistry, Faculty of Engineering, Kanagawa University, Kanagawa, Japan**
**[4]Advanced Technology Research Laboratories, Nippon Steel Corporation, Chiba, Japan**

The additive effect of noble metals to NiO-MgO solid solution on the activity and on the resistance to the carbon deposition in oxidative steam reforming of methane under pressurized condition (1.0 MPa) was investigated. It was found that the addition of small amount of Rh or Pd was effective to enhance the activity and to suppress the carbon deposition. The Pd and Rh K-edge EXAFS analysis indicated that M-Ni alloy phase (M=Rh or Pd) was formed, and FTIR spectra of CO adsorption suggests that the alloy could be formed on the surface of bimetallic particles.

## Introduction

Steam reforming of natural gas on nickel-based catalyst is the most important process for the production of hydrogen and synthesis gas (*1*) as shown by equation (1)

$$CH_4 + H_2O \rightarrow CO + 3H_2 \quad \Delta H_{298} = 206 \text{ kJ/mol} \tag{1}$$

As described above, steam reforming of methane (Eq. (1)) is a highly endothermic reaction requiring external heating process, which causes low energy efficiency. In order to improve the efficiency, the internal heating process to utilize the exothermic reactions has been examined by adding oxygen to the reactant gases. In this case, the endothermic reaction during the reforming is compensated with the exothermic one such as methane combustion (Eq. (2)) and partial oxidation of methane (Eq. (3)).

$$CH_4 + 2O_2 \rightarrow CO_2 + 2H_2O \quad \Delta H_{298} = -803 \text{ kJ/mol} \tag{2}$$
$$CH_4 + 1/2O_2 \rightarrow CO + 2H_2 \quad \Delta H_{298} = -36 \text{ kJ/mol} \tag{3}$$

The most famous process of this method is autothermal reforming of methane (ATR) developed by Haldor Topsøe (*2*), in which the exothermic and the endothermic regions are separated to avoid the hot spot formation on the catalyst surface. Here, oxygen is introduced together with the reforming agents, and this is called oxidative reforming. When the oxygen is introduced to the nickel-based reforming catalyst, the catalyst loses the reforming activity by the deep oxidation of nickel metals at the hot spot (*3-6*). Therefore, this process still needs to be improved from the viewpoint of heat exchange between the exothermic and endothermic regions. Although noble metals such as Pt and Rh are suitable for oxidative steam reforming of methane (*7, 8*), they have problems in terms of the cost and limited availability. Recently, our group has reported that the addition of small amount of Pt to Ni catalyst drastically flattens the catalyst bed temperature gradient in oxidative reforming of methane due to the inhibition of the oxidation of Ni even under the presence of oxygen. The deactivation due to oxidation can also be inhibited (*9-15*). On the other hand, it has been reported that NiO-MgO solid solution catalysts exhibited high performance in steam and dry reforming of methane (*16-22*). Especially, the catalyst had high resistance to coke deposition, which is one of the most serious problems in the reforming process. However, these catalysts are not suitable for oxidative reforming of methane because of low reducibility of Ni in the solid solution (*23*). In order to enhance the catalyst reducibility, the noble metals such as Rh and Pd were added to NiO-MgO solid solution, and the additive effect was investigated. In this study, we have focused on the additive effect of noble metals to NiO-MgO solid solution on oxidative reforming of methane at pressurized condition, especially on the suppression of coke formation which can cause catalyst deactivation.

# Experimental

## Catalyst Preparation

$Ni_{0.2}Mg_{0.8}O$ solid solution was prepared by the solid reaction method from NiO (Wako Pure Chemical Industries Ltd., Japan) and MgO (UBE Material Industries Ltd., Japan). The mixture of NiO and MgO was calcined in the air at 1423 K for 12 h. The formation of $Ni_{0.2}Mg_{0.8}O$ solid solution was identified by X-ray diffraction (XRD) as reported previously (*11*). After the calcination, the BET surface area of $Ni_{0.2}Mg_{0.8}O$ was 2.6 $m^2/g$. As a reference, MgO was calcined at the same temperature and the BET surface area was 5.7 $m^2/g$. The loading of Rh or Pd on $Ni_{0.2}Mg_{0.8}O$ and MgO was performed by the impregnation method using the aqueous solutions of $Rh(NO_3)_3$ and $Pd(NO_3)_2$. After the impregnation, the catalysts were calcined at 773 K for 3 h. The loading amount of Rh was 0.035 wt%. Pd/$Ni_{0.2}Mg_{0.8}O$ catalysts were prepared with different loading amounts of Pd from 0.1 wt% to 0.5 wt%.

## Catalytic Reaction

Oxidative steam reforming of methane was carried out in a horizontal fixed bed flow reaction system under pressurized condition. The reactor was the same with that reported previously (*10, 11*). A quartz tube (inner diameter, 4.4 mm$\phi$) inside a stainless steel tube (inner diameter, 7.4 mm$\phi$) was used as the reactor, and the quartz tube was sealed by a silicon rubber gasket. At the outlet of the catalyst bed, the reaction temperature was monitored by a thermocouple, which was inserted into a thin quartz tube (thermowell). Catalyst reduction was carried out by hydrogen flow under atmospheric pressure at 1073 K for 0.5 h in each experiment. The partial pressure ratio of reactants was $CH_4/H_2O/O_2$ = 2/1.5/1. Steam reforming ($CH_4/H_2O$=1/1) and dry reforming ($CH_4/CO_2$=1/1) of methane were also carried out in the same apparatus. The total pressure was 1.0 MPa; 30 mg catalyst was used for each experiment. The length of the catalyst bed was about 4 mm. Contact time $W/F$ was 0.43 ghmol$^{-1}$. The effluent gas was analyzed with an FID gas chromatograph (GC) (Gaskuropack 54) equipped with a methanator for $CH_4$, CO and $CO_2$. TCD–GC (Molecular Sieve 13X) was used for $H_2$ analysis.

## Characterization

Thermogravimetric analysis (TGA) for the estimation of carbon amount was carried out by using a Shimadzu DTG-60 apparatus. The analysis was done with 10 mg catalysts under flowing air (20 $cm^3$/min) at 15 K/min heating rate. The amount of carbon was estimated from the exothermic weight loss between 800 K and 1000 K.

SEM and TEM images were taken using the catalyst after the reduction (fresh) and stability test (used). SEM images were observed by S-3000 N (HITACHI). TEM images were taken by means of JEM-2010F (JEOL) equipment operated at 200 kV. The samples were dispersed in 2-propanol by supersonic waves and put on Cu grids for the TEM observation.

Rh K-edge and Pd K-edge Extended X-ray absorption fine structure (EXAFS) was measured at the BL-10B of the Photon Factory at the High Energy Accelerator Research Organization in Tsukuba, Japan. The storage ring was operated at 2.5 GeV with ring current of 300-450 mA. A Si(311) single crystal was used to obtain monochromatic X-ray beam. The structural parameters of the samples were obtained by the curve-fitting method using Rh foil and Pd foil as model compound and FEFF program (*24*). The details of the preparation of samples and the treatment of EXAFS data were described elsewhere (*25, 26*).

Fourier transform infrared (FTIR) spectra of CO adsorption were recorded with a Nicolet Magna 550 spectrometer equipped with an MCT detector (resolution: 4 $cm^{-1}$) in a transmission mode, and an *in-situ* IR quartz cell with $CaF_2$ windows was used. After the sample was pretreated in the cell, CO (0.13 kPa) was introduced, and the spectra were obtained after the evacuation.

## Results and Discussion

The effect of modification by noble metals on $Ni_{0.2}Mg_{0.8}O$ solid solution catalyst for oxidative reforming of methane ($CH_4/H_2O/O_2$ = 2/1.5/1) was examined at 973 K, 1.0 MPa total pressure and $W/F$ of 0.43 $ghmol^{-1}$. Figure 1 shows the results of $Ni_{0.2}Mg_{0.8}O$ and M/$Ni_{0.2}Mg_{0.8}O$ (M=Rh or Pd) as well as M/MgO as references. High methane conversion was obtained on $Ni_{0.2}Mg_{0.8}O$ with and without noble metal addition. This indicates that the existence of Ni is essential. Low methane conversion can be related to the low loading amount of noble metals on Rh/MgO and Pd/MgO. From the results in Figure 1, it is also found that 0.035% Rh/$Ni_{0.2}Mg_{0.8}O$ and 0.1% Pd/$Ni_{0.2}Mg_{0.8}O$ catalysts exhibited high stability.

Furthermore, we also measured the amount of deposited carbon on the catalyst surface after these activity tests. The results are shown in Figure 1-(d), and we plotted the relation to the deactivation rate over Pd/$Ni_{0.2}Mg_{0.8}O$, $Ni_{0.2}Mg_{0.8}O$ and Pd/MgO. Here, the deactivation rate is defined as follow, {($CH_4$ conv. at 0.5 h) – ($CH_4$ conv. at 8 h)}/( $CH_4$ conv. at 0.5 h) × 100. As shown in Figure 1-(d), the deactivation rate is strongly related to the amount of carbon deposition, and this suggests that the catalyst deactivation can be caused

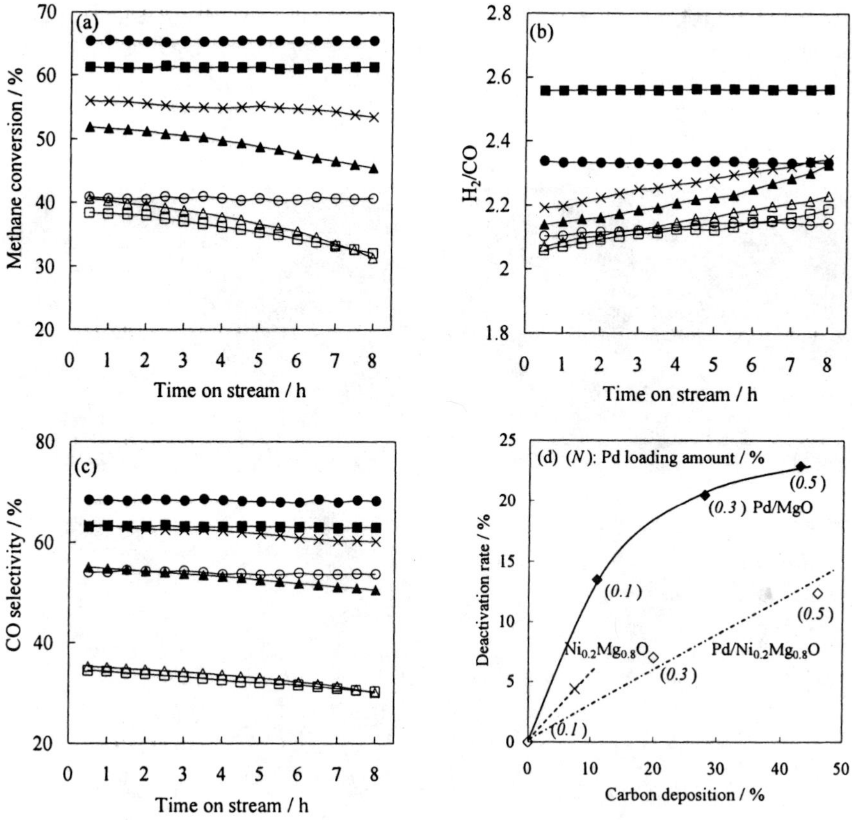


*Figure 1. Catalytic performance in oxidative reforming of methane (a-c) as a function of time on stream and relationship between carbon deposition and deactivation rate (d).*

*(×) $Ni_{0.2}Mg_{0.8}O$, (●) 0.035% Rh/$Ni_{0.2}Mg_{0.8}O$, (■) 0.1% Pd/$Ni_{0.2}Mg_{0.8}O$, (▲) 0.5% Pd/$Ni_{0.2}Mg_{0.8}O$, (○) 0.035% Rh/MgO, (□) 0.1% Pd/MgO, (Δ) 0.5% Pd/MgO. Reaction conditions: $CH_4$:$H_2O$:$O_2$ =2:1.5:1, W/F=0.43 gh/mol, catalyst weight= 0.03g, reaction temperature= 973 K, total pressure= 1.0 MPa.*

mainly by carbon deposition. It should be noted that 0.1% Pd/$Ni_{0.2}Mg_{0.8}O$ exhibited much higher resistance to carbon deposition than 0.1% Pd/MgO and $Ni_{0.2}Mg_{0.8}O$. This indicates the synergy between Pd and $Ni_{0.2}Mg_{0.8}O$ in resistance to carbon deposition. In addition, no carbon was formed on Rh/ $Ni_{0.2}Mg_{0.8}O$ and Rh/MgO. In the case of Pd/ $Ni_{0.2}Mg_{0.8}O$, the loading amount of Pd can strongly influence the synergistic effect, and 0.1% was more effective than 0.3 and 0.5%.

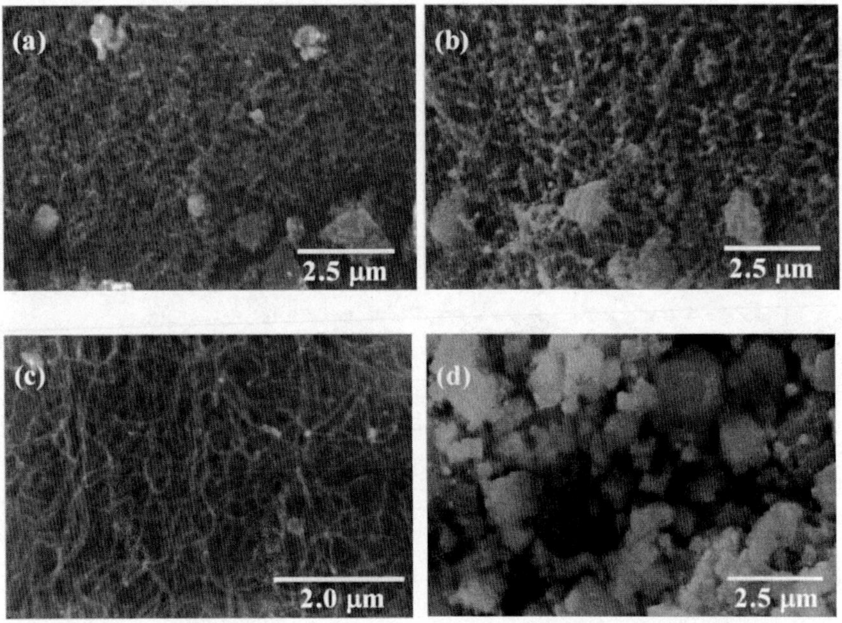


*Figure 2. SEM images of catalysts after stability test. (a) 0.5% Pd/MgO, (b) 0.5% Pd/ $Ni_{0.2}Mg_{0.8}O$, (c) $Ni_{0.2}Mg_{0.8}O$, (d) 0.1% Pd/ $Ni_{0.2}Mg_{0.8}O$. Stability test conditions are described in Figure 1.*

Over Pd/MgO catalysts, large amount of carbon was deposited during oxidative steam reforming of methane at 1.0 MPa, and the amount increased with increasing loading amount of Pd on MgO. And almost the same amount of deposited carbon was obtained over 0.5% Pd/$Ni_{0.2}Mg_{0.8}O$. As shown in Figure 2-(a) and Figure 2-(b), it was found that the most part of catalyst surface of 0.5% Pd/MgO and 0.5% Pd/$Ni_{0.2}Mg_{0.8}O$ after the activity test was covered with large amount of deposited carbon. Although TEM observation of used 0.5% Pd/$Ni_{0.2}Mg_{0.8}O$ was also carried out, the image was not clear due to too much deposited carbon as shown in Figure 3-(d). The metal particles were hardly recognized because they were embedded in the carbon.

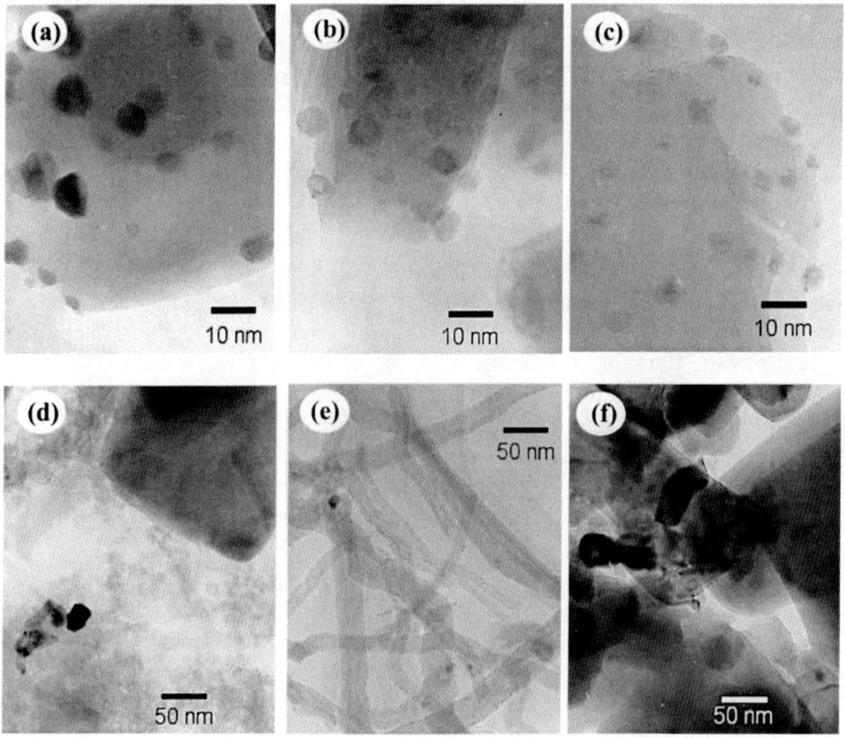


*Figure 3. TEM images of catalysts after reduction (a ,b ,c) and stability test. (d, e, f). (a, d) 0.5% Pd/$Ni_{0.2}Mg_{0.8}O$, (b, e) $Ni_{0.2}Mg_{0.8}O$, (c, f) 0.1% Pd/$Ni_{0.2}Mg_{0.8}O$. Stability test conditions are described in Figure 1.*

The carbon deposition on $Ni_{0.2}Mg_{0.8}O$ was as large as 7wt% after the stability test of oxidative reforming of methane (Figure 1-(d)). The SEM image of used $Ni_{0.2}Mg_{0.8}O$ is shown in Figure 2-(c). The morphology of deposited carbon on $Ni_{0.2}Mg_{0.8}O$ seems to be different from that of 0.5% Pd/MgO and 0.5% Pd/$Ni_{0.2}Mg_{0.8}O$. The carbon deposited on $Ni_{0.2}Mg_{0.8}O$ seems to be so-called carbon-fibers, being thinner and longer than the carbon on 0.5% Pd/NiMgO. From the TEM observation (Figure 3-(e)), it was easy to find the carbon-fibers possessing the aggregated metal particles on the tip. And most of carbon-fibers seemed to be hollow.

On the other hand, 0.1% Pd/$Ni_{0.2}Mg_{0.8}O$ showed remarkably high resistance to carbon deposition. Even after the stability test at 1.0 MPa for 8h, the amount of deposited carbon was estimated to be almost zero, as shown in Figure 1-(d). This result is also supported by the SEM and TEM observation, as shown in Figure 2-(d) and Figure 3-(f), respectively. Most part of the catalyst surface was still clean without carbon deposition after the stability test. Figures 3 (a-c) show the TEM images of the fresh catalysts. Average particle size (*27*) can be estimated to be 7.6±0.6, 6.2±0.5 and 4.7±0.5 nm over 0.5% Pd/$Ni_{0.2}Mg_{0.8}O$, $Ni_{0.2}Mg_{0.8}O$ and 0.1% Pd/$Ni_{0.2}Mg_{0.8}O$, respectively. The difference in metal particle size is so small, and the resistance to carbon deposition can not be explained by the difference in the metal particle size.

Figure 4 shows catalytic performance of dry and steam reforming of methane over $Ni_{0.2}Mg_{0.8}O$, 0.1% Pd/$Ni_{0.2}Mg_{0.8}O$ and 0.1% Pd/MgO. In both cases, the order of methane conversion was as follows; 0.1% Pd/$Ni_{0.2}Mg_{0.8}O$ > $Ni_{0.2}Mg_{0.8}O$ > 0.1% Pd/MgO. This behavior is almost similar to that of oxidative steam reforming of methane. On the other hand, methane conversion decreased gradually with time on stream in both reactions over these three catalysts, although the stable activity was obtained over 0.1% Pd/$Ni_{0.2}Mg_{0.8}O$ during oxidative steam reforming of methane.

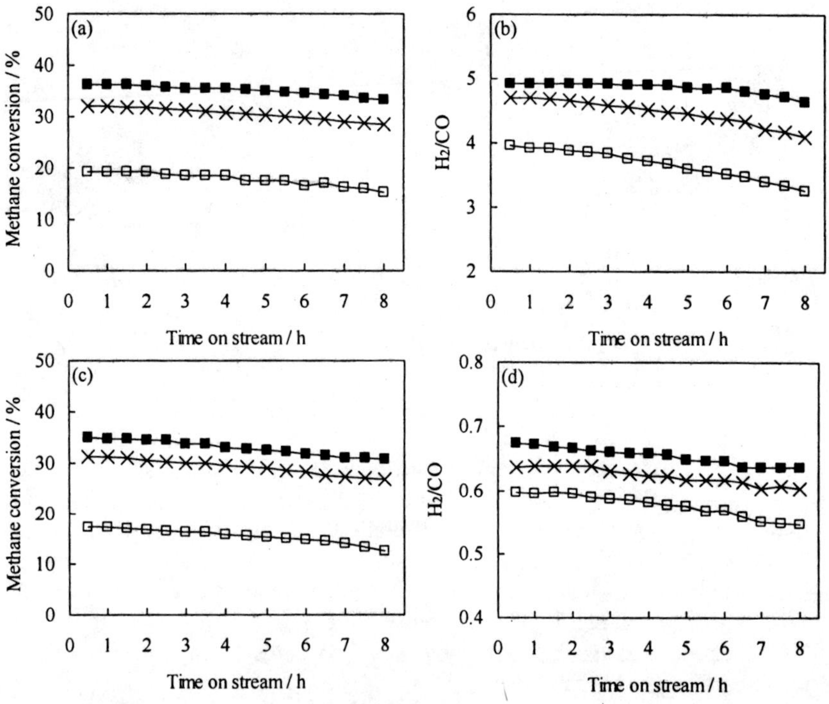


*Figure 4. Catalytic activity and stability for steam and dry reforming of methane. (a, b) $CH_4/H_2O=1/1$, (c, d) $CH_4/CO_2=1/1$. (×) $Ni_{0.2}Mg_{0.8}O$, (■) 0.1% Pd/$Ni_{0.2}Mg_{0.8}O$, (□) 0.1% Pd/MgO. Reaction conditions: catalyst weight: 0.03 g, reaction temperature 973 K, total pressure: 1.0 MPa, W/F: 0.43 ghmol$^{-1}$.*

Figure 5 shows the relationship between the carbon deposition and deactivation rate in steam, dry and the oxidative steam reforming. This observation suggests that the deactivation was strongly related to the coke formation on the catalyst in every method of reforming. The amount of carbon deposition was in the order of dry, steam and the oxidative reforming of methane. $Ni_{0.2}Mg_{0.8}O$ has showed the highest resistance to the carbon deposition, but under pressurized condition the coke formation on $Ni_{0.2}Mg_{0.8}O$ was as large as that on 0.1% Pd/MgO. Although 0.1% Pd/$Ni_{0.2}Mg_{0.8}O$ catalyst was able to reduce the carbon deposition in dry and steam reforming, it was still as large as 10wt%. On the other hand, 0.1% Pd/$Ni_{0.2}Mg_{0.8}O$ has a possibility to suppress carbon deposition completely in oxidative reforming of methane, even under pressurized condition.

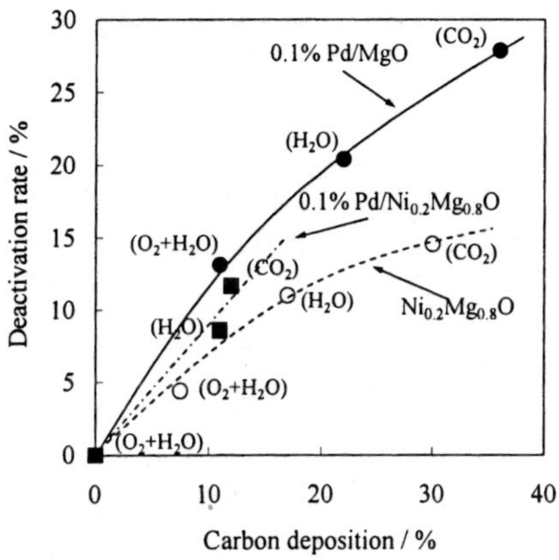


*Figure 5. Relation between carbon deposition and deactivation rate in the reaction tests of steam ($H_2O$), dry ($CO_2$) and the oxidative steam ($O_2$+$H_2O$) reforming of methane over 0.1% Pd/$Ni_{0.2}Mg_{0.8}O$ (■), $Ni_{0.2}Mg_{0.8}O$ (○) and 0.1% Pd/MgO (●). Reaction conditions are described in Figures 1 and 5.*

## Structural Characterization of Modified NiO-MgO by Small Amount of Noble Metals using EXAFS and CO-FTIR

Figure 6 and Table 1 show the results of EXAFS spectra of MgO and $Ni_{0.2}Mg_{0.8}O$ supported Rh and Pd catalysts after hydrogen reduction. Based on the results mentioned above, the catalysts with lower loading amount of noble metal are more important. However, the loading amount is too small for the EXAFS measurement. Therefore, we showed the results of 0.5 wt% catalysts.

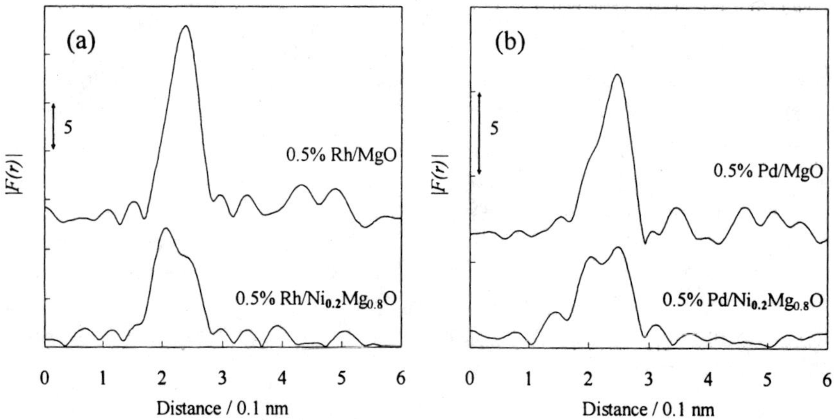


*Figure 6. Fourier transform of $k^3$-weighted Rh K-edge (a) and Pd K-edge (b) EXAFS of reduced catalysts.*
*(a) FT range = 35-127 $nm^{-1}$, (b) FT range = 35-124 $nm^{-1}$.*

**Table 1. Curve Fitting Results of Rh and Pd K-edge EXAFS of 0.5% Rh and Pd Catalysts after Reduction at 1073 K**

| *Samples* | *Shells* | $CN^a$ | $R^b$ / $10^{-1}$ nm | $\sigma^c$ / $10^{-1}$ nm | $\Delta E_0{}^d$ / eV | $R_f{}^e$ / % |
|---|---|---|---|---|---|---|
| 0.5% Rh/MgO | Rh-Rh | 9.7 | 2.67 | 0.076 | 2.1 | 0.51 |
| 0.5% Rh/$Ni_{0.2}Mg_{0.8}O$ | Rh-Rh | 7.4 | 2.65 | 0.093 | 9.6 | |
| | Rh-Ni | 3.8 | 2.51 | 0.087 | -8.1 | 0.48 |
| Rh foil | Rh-Rh | 12 | 2.68 | 0.060 | | |
| 0.5% Pd/MgO | Pd-Pd | 8.0 | 2.72 | 0.088 | -0.7 | 1.0 |
| 0.5% Pd/$Ni_{0.2}Mg_{0.8}O$ | Pd-Pd | 8.6 | 2.71 | 0.114 | 6.2 | 1.0 |
| | Pd-Ni | 2.7 | 2.53 | 0.118 | -8.6 | |
| Pd foil | Pd-Pd | 12 | 2.72 | 0.060 | 0.0 | |

[a]Coordination number, [b]Bond distance, [c]Debye-Waller factor, [d]Difference in the origin of photoelectron energy between the reference and the sample, [e]Residual factor, Fourier transform range: 35-127 $nm^{-1}$ for Rh and 35-124 $nm^{-1}$ for Pd, Fourier filtering range: 0.160-0.280 nm for Rh and 0.170-0.295 nm for Pd.

In the case of Pd/MgO and Rh/MgO, good fitting was obtained by only the Pd-Pd and Rh-Rh bond. In contrast, the contribution of Pd-Ni or Rh-Ni bond as well as Pd-Pd and Rh-Rh bond was observed. The bond length of Pd-Ni and Rh-Ni was close to that of the Ni-Ni bond (0.249 nm) in the Ni metal phase. This suggests the formation of Pd-Ni and Rh-Ni alloy with high Ni content. In addition, the coordination number of Rh-Ni on Rh/$Ni_{0.2}Mg_{0.8}O$ was larger than that of Pd-Ni on Pd/$Ni_{0.2}Mg_{0.8}O$. This suggests that the interaction of Rh-Ni is stronger than Pd-Ni.

In order to evaluate the surface properties of the metal particles, we measured the FTIR spectra of CO adsorption on reduced fresh catalysts as shown in Figure 7. On $Ni_{0.2}Mg_{0.8}O$ catalyst, two peaks at 2050 $cm^{-1}$ and 1930 $cm^{-1}$ due to linear and bridge CO species (*18, 28-30*) were observed. It is characteristic that the intensity of bridge CO is relatively strong compared to that of linear CO. The addition of Rh changed the CO adsorption profile even when the additive amount of Rh is rather small. On 0.035% $Rh/Ni_{0.2}Mg_{0.8}O$, the peak due to linear CO at 2050 $cm^{-1}$ became broad and it seems that two peaks are overlaped, one is linear CO on Ni and the other is linear CO on Rh (*31-34*). The peak due to bridge CO at 1938 $cm^{-1}$ became weaker than that due to linear CO. These results suggest that Rh atoms are located on the surface of metal particles even though the Rh amount is very small.

In the case of $Pd/Ni_{0.2}Mg_{0.8}O$, a new peak at 1980 $cm^{-1}$ appeared. As a reference, the spectrum of 0.5% Pd/MgO is also shown in Figure 7. According to the previous reports (*35-40*), the peak at 1970 $cm^{-1}$ is assigned to bridge CO on Pd and the peak at 1900 $cm^{-1}$ is assigned to CO adsorbed on the 3-fold hollow site of Pd surface. From the comparison of spectra between Pd/MgO and $Pd/Ni_{0.2}Mg_{0.8}O$, the peak at 1980 $cm^{-1}$ can be assigned to bridge CO on Pd. This indicates the presence of Pd on the surface of metal particles, and the peak shift can be related to the formation of Pd-Ni alloy.

It is possible to determine the ratio of noble metal to metallic nickel ($M/Ni^0$, M=Rh and Pd) from the amount of $H_2$ consumption in the temperature-programmed reduction profile, and it was in the range of 0.01 to 0.05 (*10*).

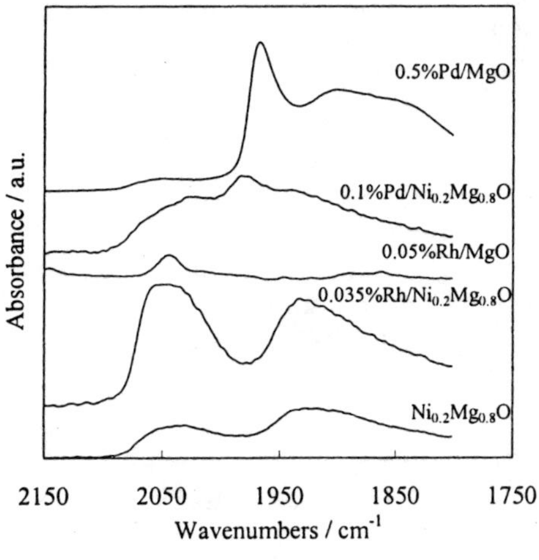


*Figure 7. FTIR spectra of CO adsorption. CO was evacuated before measurement after adsorption at 0.13 kPa at room temperature.*

If this average composition of noble metals in M-Ni alloy is true for the surface composition of M-Ni alloy particles, it is too low to obtain the drastic change in FTIR spectra as shown and discussed in Figure 7. Therefore, it can be concluded that the surface composition of $M/Ni^0$ was much higher than the average one, probably because of surface segregation of noble metals in M-Ni alloy particles. This surface M-Ni alloy phase is suggested to cause the high performance of the oxidative methane reforming of methane.

## Conclusions

Modification of NiO-MgO solid solution by small amount of noble metals, such as Rh and Pd was effective to improve the catalytic activity for oxidative reforming of methane under pressurized condition. In the case of Pd addition, the loading amount was very important and excessive addition of Pd had negative effect on the activity and stability. Especially, it was found that the deactivation rate can be strongly related to the amount of carbon deposition. The tendency of steam and dry reforming of methane was similar to that of oxidative reforming of methane. It should be noted that the catalyst made of 0.1% $Pd/Ni_{0.2}Mg_{0.8}O$ showed quite high stability and resistance to coke formation even under pressurized condition, while 0.5% $Pd/Ni_{0.2}Mg_{0.8}O$ exhibited lower activity and stability than NiO-MgO itself and promoted the carbon formation. From the structural analysis using EXAFS and FTIR, it was evident that the alloy phase can be formed between Ni and noble metals (Rh and Pd) and located mainly on the surface of metal particles. This alloy phase and its composition on the surface is considered to influence strongly the catalytic performance and stability of the modified catalysts.

## Acknowledgements

A part of this study was supported by the Industrial Technology Research Grant Program (05A43002C) from the New Energy and Industrial Technology Development Organization (NEDO) of Japan. EXAFS studies were performed under the approval of the Photon Factory Advisory Committee (Proposal No. 2006G095).

## References

1. Rostrup-Nielsen, J.R. *Catal. Rev. Sci. Eng.* **2004**, *46*, 247.
2. Rostrup-Nielsen, J.R. *Sekiyu Gakkaishi* **1997**, *40*, 366.
3. Li, B.; Maruyama, K.; Nurunnabi, M.; Kunimori, K.; Tomishige, K. *Ind. Eng. Chem. Res.* **2005**, *44*, 485.

4. Li, B.; Watanabe, R.; Maruyama, K.; Nurunnabi, M.; Kunimori, K.; Tomishige, K. *Appl. Catal. A Gen.* **2005**, *290*, 36.
5. Li, B.; Watanabe, R.; Maruyama, K.; Kunimori, K.; Tomishige, K. *Catal. Today* **2005**, *104*, 7.
6. Tomishige, K.; Nurunnabi, M.; Maruyama, K.; Kunimori, K. *Fuel Process. Technol.* **2004**, *85*, 1103.
7. Tomishige, K.; Kanazawa, S.; Suzuki, K.; Asadullah, M.; Sato, M.; Ikushima, K.; Kunimori, K. *Appl. Catal. A Gen.* **2002**, *233*, 35.
8. Li, B.; Maruyama, K.; Nurunnabi, M.; Kunimori, K.; Tomishige, K. *Appl. Catal. A Gen.* **2004**, *275*, 157.
9. Nagaoka, K.; Jentys, A.; Lercher, J.A. *J. Catal.* **2005**, *229*, 185.
10. Nurunnabi, M.; Mukainakano, Y.; Kado, S.; Li, B.; Kunimori, K.; Suzuki, K.; Fujimoto, K.; Tomishige, K. *Appl. Catal. A Gen.* **2006**, *299*, 145.
11. Nurunnabi, M.; Li, B.; Kunimori, K.; Suzuki, K.; Fujimoto, K.; Tomishige, K. *Appl. Catal. A Gen.* **2005**, *292*, 272.
12. Nurunnabi, M.; Fujimoto, K.; Suzuki, K.; Li, B.; Kado, S.; Kunimori, K.; Tomishige, K. *Catal. Commun.* **2006**, *7*, 73.
13. Nurunnabi, M.; Li, B.; Kunimori, K.; Suzuki, K.; Fujimoto, K.; Tomishige, K. *Appl. Catal. Lett.* **2005**, *103*, 277.
14. Tomishige, K.; Kanazawa, S.; Ito, S.; Kunimori, K. *Appl. Catal. A Gen.* **2003**, *244*, 71.
15. Tomishige, K.; Kanazawa, S.; Sato, M.; Ikushima, K.; Kunimori, K. *Catal. Lett.* **2002**, *84*, 69.
16. Yamazaki, O.; Tomishige, K.; Fujimoto, K. *Appl. Catal. A Gen.* **1996**, *136*, 49.
17. Tomishige, K.; Chen, Y.G.; Fujimoto, K. *J. Catal.* **1999**, *181*, 91.
18. Chen, Y.G.; Tomishige, K.; Yokoyama, K.; Fujimoto, K. *J. Catal.* **1999**, *184*, 479.
19. Hu, Y.H.; Ruckenstein, E. *Catal. Rev. Sci. Eng.* **2002**, *44*, 423.
20. Tomishige, K.; Himeno, Y.; Matsuo, Y.; Yoshinaga, Y.; Fujimoto, K. *Ind. Eng. Chem. Res.* **2000**, *39*, 1891.
21. Tomishige, K. *Catal. Today* **2004**, *89*, 405.
22. Tomishige, K.; Matsuo, Y.; Yoshinaga, Y.; Sekine, Y.; Asadullah, M.; Fujimoto, K. *Appl. Catal. A Gen.* **2002**, *223*, 225.
23. Tomishige, K.; Matsuo, Y.; Yoshinaga, Y.; Asadullah, M.; Sekine, Y.; Fujimoto, K. *ACS Symp. Ser.* **2002**, *809*, 303.
24. Ankudinov, A.L.; Ravel, B.; Rehr, J.J.; Conradson, S.D. *Phys. Rev. B* **1998**, *58*, 7565.
25. Okumura, K.; Amano, J.; Yasunobu, N.; Niwa, M. *J. Phys. Chem. B* **2000**, *104*, 1050.
26. Okumura, K.; Matsumoto, S.; Nishiaki, N.; Niwa, M. *Appl. Catal. B Environ.* **2003**, *40*, 151.

27. Chen, Y.G.; Tomishige, K.; Yokoyama, K.; Fujimoto, K. *Appl. Catal. A Gen.* **1997**, *165*, 335.
28. Peri, J.B. *J. Catal.* **1984**, *86*, 84.
29. Bartholomew, C.H.; Pannell, R.B.; Butler, J.L. *J. Catal.* **1980**, *65*, 335.
30. Zecchina, A.; Spoto, G.; Coluccia, S.; Guglielminotti, E. *J. Chem. Soc. Faraday Trans.* **1984**, *80*, 1875.
31. Yang, A.Y.; Garland, C.W. *Phys. Chem.* **1957**, *61*, 1505.
32. Yates, Jr., J.T.; Duncan, T.M.; Worley, S.D.; Vaughan, R.W. *J. Chem. Phys.* **1979**, *70*, 1219.
33. Cavanagh, R.R.; Yates, Jr., J.T. *J. Chem. Phys.* **1981**, *74*, 4150.
34. Solymosi, F.; Pasztor, M. *J. Am. Chem. Soc.* **1985**, *89*, 4789.
35. Bertarione, S.; Scarano, D.; Zecchina, A.; Johanek, V.; Hoffmann, J.; Schauermann, S.; Frank, M.M.; Libuda, J.; Rupprechter, G.; Freund, H.J. *J. Phys. Chem. B* **2004**, *108*, 3603.
36. Liotta, L.F.; Martin, G.A.; Deganello, G. *J. Catal.* **1996**, *164*, 322.
37. Sheu, L.L.; Karpinski, Z.; Sachtler, W.M.H. *J. Phys. Chem.* **1989**, *93*, 4890.
38. Bradshaw, A.M.; Hoffmann, F.M. *Surf. Sci.* **1978**, *72*, 513.
39. Ortega, A.; Hoffmann, F.M.; Bradshaw, A.M. *Surf. Sci.* **1982**, *119*, 79.
40. Hoffmann, F.M. *Surf. Sci. Report* **1983**, *3*, 107.

# Production from Natural Gas

# Chapter 7

# Multifunctional Catalyst for Fischer–Tropsch Synthesis

**Mitrajit Mukherjee and Sankaran Sundaresan**

**Exelus Inc., 99 Dorsa Avenue, Livingston, NJ 07039**

A first-of-a-kind multifunctional catalytic system is being developed to convert synthesis gas into synthetic crude via the Fischer-Tropsch reaction. This is achieved through intensification of chemical reaction and heat and mass transport processes within the catalyst system. The synergistic integration of intensified unit operations with chemical reaction leads to enhanced catalyst performance and significant economic advantages.

Gas-To-Liquids (GTL) technologies offer opportunities to convert natural gas into premium transportation grade fuels. The GTL process first converts natural gas to synthesis gas (a mixture of CO & $H_2$) by partial oxidation of methane, steam reforming of methane or a combination of both called auto-thermal reforming. Syngas (synthesis gas) is then upgraded into a range of products termed 'synthetic crude' via the Fischer-Tropsch (FT) reaction. The major advantage of the GTL process is the exceptional quality of its products, which can easily meet current and future anticipated clean fuels specifications.

The initial gas-to-liquids complexes implemented in the 1980s and early 1990s were not commercially successful for a number of reasons – the main one being that they were far too expensive. Although improvements in technology have been achieved during the past decade, GTL complexes are still very expensive propositions. The other problem with GTL projects is scalability. The lowest capacity of the three commercial GTL processes available requires throughput of at least $100 \times 10^6$ $ft^3$/day of natural gas to be within its viable range and have minimum estimated installed costs of about $1 billion for a 34,000 barrels/day facility (*1*). This paper summarizes the development of a break-through FT technology that has the ability to significantly improve the economics of the process using a multi-functional catalytic system. The cost reduction potential using this next-generation FT technology provides a very strong springboard for overall GTL process viability.

## Fischer-Tropsch Process Technical Challenges

The Fischer-Tropsch process presents many technical challenges due to its unique reaction chemistry and high heats of reaction. The selectivity of the process is usually described by the Anderson–Schulz–Flory (ASF) distribution and the characteristic chain growth probability, $\alpha$. High chain growth probabilities ($\alpha$ >0.90) are regarded desirable as they lead to high selectivities of heavy components (*2*), which subsequently can be converted to liquid fuels rather easily. Product distribution or product selectivity ($\alpha$) depends heavily on the type and structure of the catalysts and on the reactor type and operating conditions.

### FT Catalyst Design Issues

The most common FT catalysts are Group VIII metals (Co, Fe, Ru, etc.) (*3*). Most early FT catalysts were Fe-based systems mainly due to its low cost. Cobalt catalysts give the highest yields, longest lifetimes and yield predominantly linear alkanes. A large number of papers have focused on the optimization of metal. However limitations in pore diffusion within the support lead to lower $C_{5+}$ product selectivity. Even though the reactants are in the gas phase, the pores of the catalyst are filled with liquid products. The diffusion rates in the liquid phase are typically three orders of magnitude slower than in the gas phase, and even slow reactions may be diffusion limited in the liquid phase. As the ratio between hydrogen and carbon monoxide concentration is important, the difference in diffusivity between hydrogen and carbon monoxide result in selectivity changes. Increasing the transport limitations leads to CO depletion and enhanced hydrogenation reactions, resulting in lower selectivities to $C_{5+}$. Hence, key to an effective FT catalyst design lies in optimizing both the catalyst composition and its diffusional characteristics.

**FT Reactor Design Issues**

The FT synthesis reaction is strongly exothermic, requiring highly effective heat transfer. Multi-tubular fixed-bed and slurry bubble columns (*4*) have been the reactors of choice for low-temperature Fischer–Tropsch synthesis. The large support particles in fixed-bed reactors result in poor intra-particle mass transfer characteristics and the space–time yield is limited by heat transfer in the catalyst bed. The slurry system gives rise to significantly improved mass transfer characteristics within the catalyst particles, but the separation of the catalyst from the product can be troublesome. Back-mixing renders the slurry reactor less efficient in terms of reactor volume than a plug flow reactor. A principal advantage of slurry bubble column reactors (BCR) over fixed-bed reactors is that the presence of a circulating/agitated slurry phase greatly increases the heat transfer rate to cooling surfaces built into the reactor. As a result, conventional reactors are unable to satisfy all requirements simultaneously.

## New Development Approach

In commercial-scale FT reactors, non-idealities in flow distribution and inefficient inter-phase transport rates reduce the effectiveness of the catalyst significantly. As a result, large improvements in catalyst performance alone do not translate to large improvements in reactor performance.

Researchers at Exelus are developing a first-of-a-kind catalytic system for multi-phase reactions that blurs the line of distinction between the reactor and catalyst by incorporating many features one associates with a reactor within the catalyst body. The new multifunctional catalyst system, called the HyperCat-FT, has a controlled structure at scales ranging from the nano- (pore level) to the macro-scale (reactor level). By creating 'structure' on various length scales, the HyperCat-FT system is able to achieve superior control of the micro-environment at the catalytic sites through higher rates of gas-liquid and liquid-solid heat and mass transfer while minimizing pore-diffusion limitations. The high level of control allows ideal reaction conditions to prevail at the active sites, enhancing catalyst performance leading to significantly higher activity and product selectivity. As a result, reactor productivity is enhanced and the active catalyst volume can now be reduced by 50% or more over a conventional slurry bubble column reactor. The higher levels of inter-phase heat transfer also allow steam coils (a low-cost item) to be used for removing reaction heat instead of expensive shell-and-tube reactor designs. These characteristics result in a significant reduction in the cost of the Fischer-Tropsch reactor, leading to a corresponding decrease in the cost of a GTL complex.

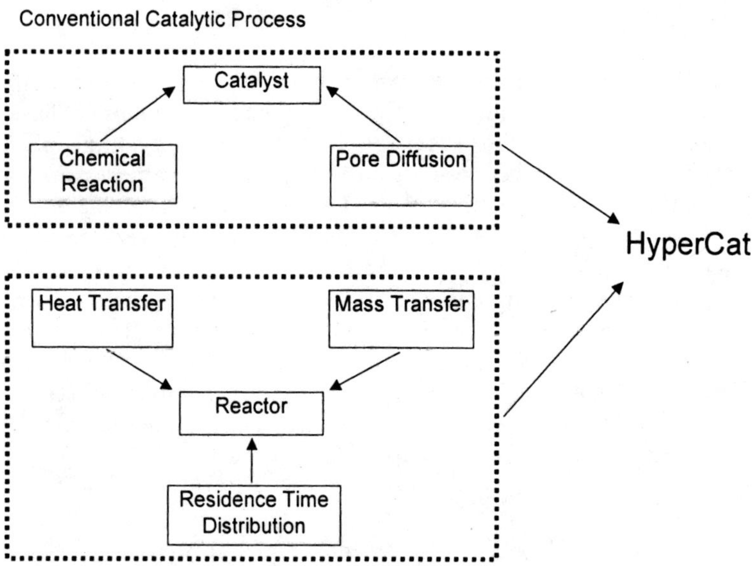


*Figure 1. Concept for a new multifunctional FT-catalyst.*

## Results and Discussion

### Gas Hold-Up

Gas hold-up is a critical parameter in characterizing the hydrodynamic behavior and hence the performance of a bubble column reactor. It determines: a) the reaction rate by controlling the gas-phase residence time and b) the mass-transfer rate by governing the gas-liquid interfacial area. It is mainly a function of the gas velocity and the liquid physical properties.

The gas hold-up of the HyperCat system was studied using an air-water system. The gas superficial velocities were varied from 0.1-1.0 m/s. The results are shown in Figure 2. The HyperCat gas hold-up is slightly lower than that in a conventional bubble column reactor (~ 30% reduction) and comparable to that in a slurry bubble column with 30-35% solids concentration (*4*).

### Liquid-Phase Axial Dispersion

The liquid phase axial dispersion in bubble column reactors is very high making them effectively back-mixed reactors for industrial-scale systems.

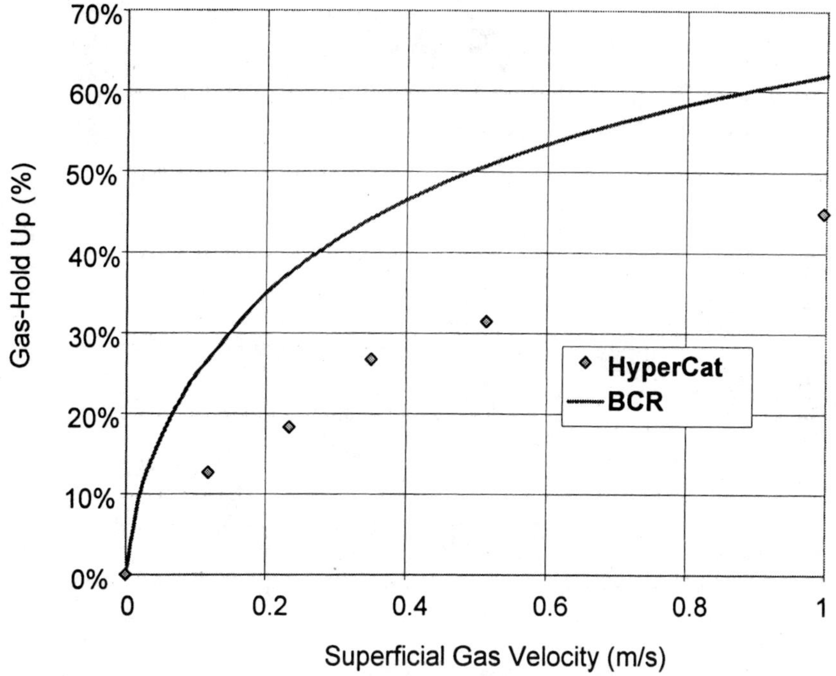


*Figure 2. Gas hold-up – HyperCat versus empty bubble column.*

Liquid mixing in bubble columns is a result of global convective re-circulation of the liquid phase and turbulent diffusion due to eddies generated by the rising bubbles. By structuring the gas and the liquid flow, HyperCat reduces the axial dispersion for both phases leading to a large reduction in axial dispersion coefficients ($D_{AX}$). The real benefit of this reduced dispersion is that it is not a function of the column diameter ($D_C$) as is the case with conventional bubble column reactors.

We have measured the liquid phase $D_{AX}$ for a range of gas velocities ($U_G$) shown in Figure 3. The dispersion values measured for the HyperCat are more than two orders of magnitude lower than those found in commercial scale reactors – which means that the HyperCat reactor will behave like a plug-flow reactor for both the liquid and gas-phases. Dekwer (*5*) previously estimated commercial bubble column performance.

## Heat Transfer

The hydrocarbon product spectrum produced by a FT catalyst is highly dependent upon catalyst temperature. The reaction is highly exothermic and, if rates of heat removal from the catalyst are not sufficiently high, "hot-spots" will

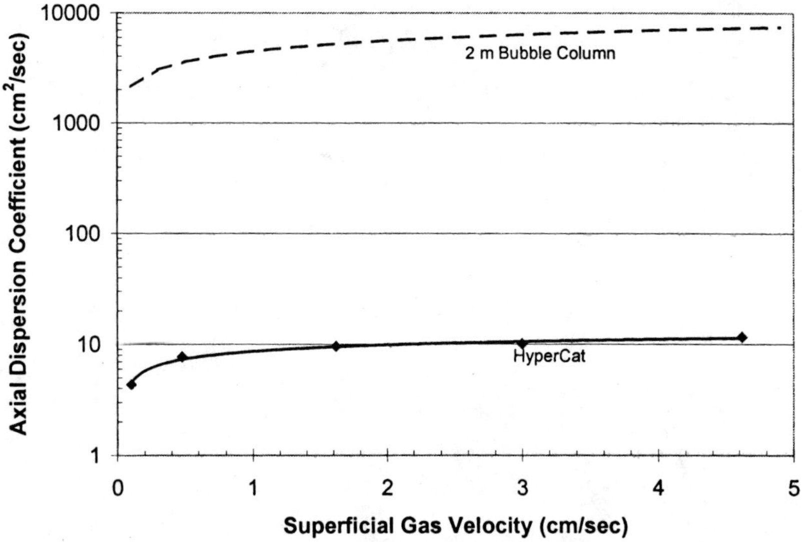


*Figure 3. Liquid phase axial dispersion in HyperCat and Bubble Columns*

form which will result in degradation of the product. Also, if the catalyst is exposed to too high a temperature, carbon will be formed, which may damage the structural integrity of the catalyst.

The HyperCat's unique design enables it to achieve high rates of heat-transfer despite a low liquid phase axial dispersion. In addition, the presence of gas bubbles in the heat-transfer zone allows continuous replacement of the liquid film in contact with the heat-exchanger surface, leading to heat transfer rates which are similar to that obtained in a bubble column reactor under comparable gas velocities. However, since the HyperCat system will operate at higher gas velocities than possible in a slurry bubble column system, the heat transfer coefficient will be higher for the HyperCat system. The heat transfer coefficient of the HyperCat system was measured using an air-water system and the results are shown in Figure 4. To put the results in perspective, the HyperCat modules were removed from the reactor and the heat transfer coefficient were measured for a conventional bubble column reactor without any solids.

The data clearly shows that the heat transfer from the bulk liquid to the walls of the heat-transfer tubes is about 15-20% higher for the HyperCat as compared to a conventional bubble column reactor.

## Catalyst Performance

A major stumbling block in converting structured carriers into structured catalysts is that for the same amount of active catalyst loading, structured

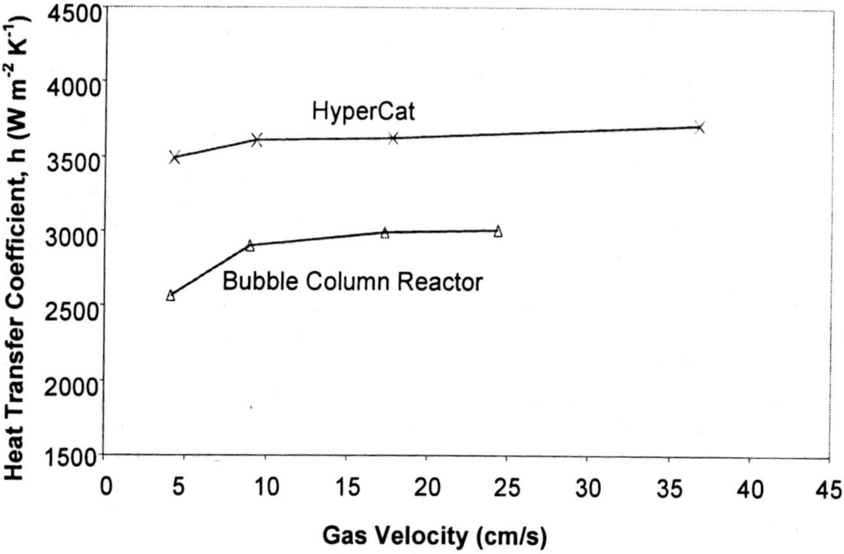


*Figure 4. Heat-transfer characteristics of the HyperCat system.*

catalysts give much lower activities than traditional powder catalysts. To compare performance of the powder catalyst and the HyperCat on an equivalent cobalt weight basis we used the data from the powder catalyst to generate CO conversion for the given condition as a function of the Damkohler number ($k\tau$). The data from the HyperCat was then used to plot CO conversion as a function of "equivalent Damkohler number" which corrects for the slightly lower cobalt loading for the HyperCat-FT system (Figure 5).

It is very clear from the data above that there is virtually no difference in the performance of the HyperCat as compared to the powder catalyst. This result is extremely important in light of other efforts being made to develop structured catalysts for Fischer-Tropsch reaction, which show that the activity of the structured catalyst is always lower than the powder catalyst (*6*).

## Comparison with slurry bubble column reactor

Finally, we studied the effect of liquid dispersion on catalyst performance by comparing the performance of the powder catalyst in a bubble column reactor with the HyperCat-FT system. As shown in Figure 6, the CO conversion is much lower for a low Peclet number (bubble column reactor with a back-mixed liquid phase) as opposed to a higher Peclet number for the HyperCat system. The tests were conducted under the same process conditions and Damkohler number. The change in Peclet number did not change the liquid product

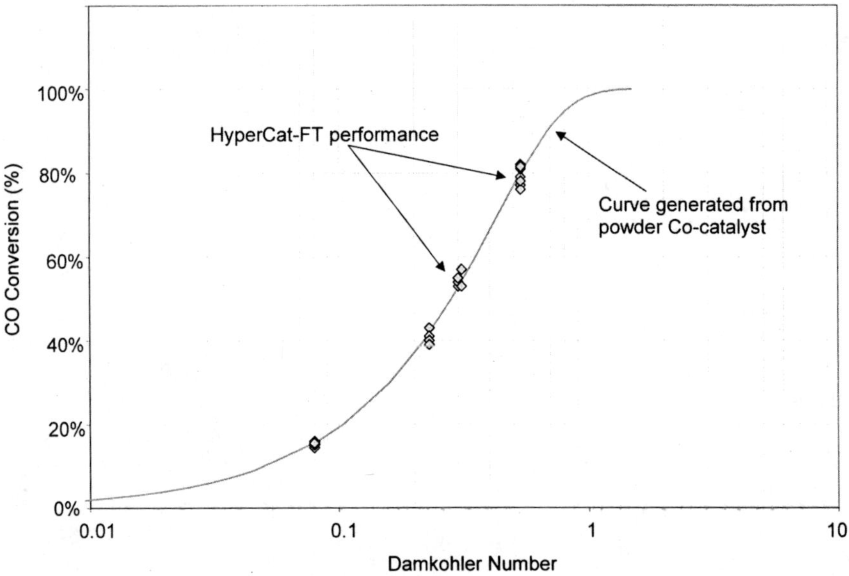


*Figure 5. HyperCat-FT catalyst performance compared to powder FT catalyst.*

distribution or the methane selectivity – which appear to be mainly a function of reaction temperature and pressure.

## Comparison with current FT processes

The Fischer-Tropsch reactor design challenge can be distilled down to the following critical issues:

1. Achieve a high conversion per pass, in order to achieve high yields and reduce recycle and investment costs.
2. Eliminate the solids handling issue.
3. Provide high heat-transfer rates to remove the large reaction heat.
4. Minimize pore and film diffusion limitations (*7*).

As seen from the above analysis, issues 1 and 2 can be easily met in a fixed-bed reactor, while issues 3 and 4 are more suitable in a back-mixed bubble column reactor. However, one reactor alone is unable to satisfy all four requirements simultaneously.

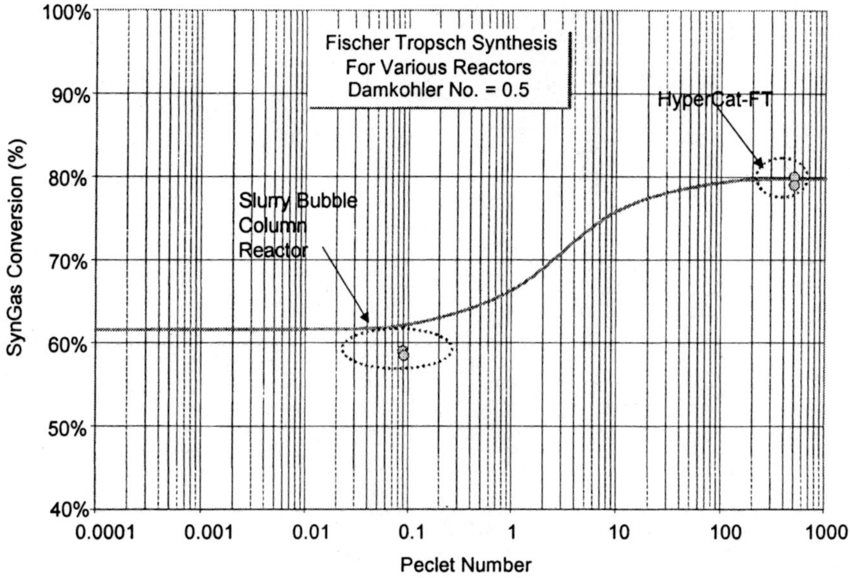


*Figure 6. Effect of liquid phase axial dispersion on CO conversion.*

As shown in Figure 7, HyperCat integrates the inter-phase transport advantages of slurry operation, the scalability of fixed-bed operation and the high heat removal ability of liquid-phase operation in one neat package. We believe that the synergistic combination of these characteristics will allow development of a practical, cost-effective 3rd generation FT technology that cuts investment costs by ~ 35%.

## Conclusions

Exelus has developed a novel structured catalytic system that allows one to meet all four criteria in a single catalytic system. Hydrodynamic tests reveal that the HyperCat has similar gas hold-up as a slurry bubble column reactor but with a much lower liquid axial-dispersion coefficient. Cold-flow studies appear to indicate that the heat-transfer coefficient of this new system is similar to a bubble column reactor. Catalyst performance tests reveal that the performance of the HyperCat is similar to that of a powder catalyst when used in a plug-flow reactor.

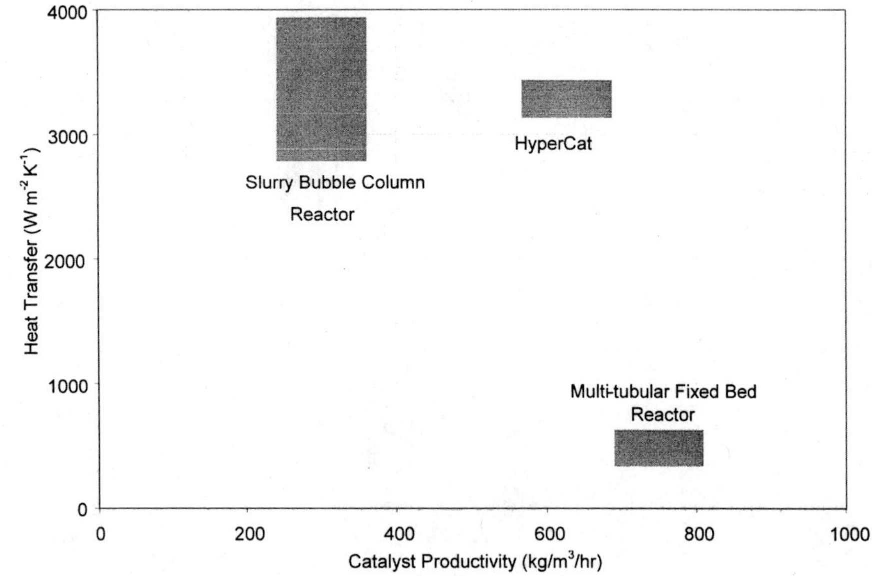


*Figure 7. Characteristics of conventional FT reactors compared with HyperCat.*

## Acknowledgement

This work was funded by grants from the U.S. Department of Energy.

## References

1. Dry, M.E., The Fischer–Tropsch process: 1950–2000, *Catalysis Today*, **2002**, 71.
2. De Swart, J.W.A.; Krishna, R.; Sie, S.T., Selection, design and scale up of the Fischer-Tropsch reactor, *Stud. Surf. Sci. Catal.* **1997**, 107, pp. 213–218.
3. Burtron H. Davis, "Overview of reactors for liquid phase Fischer–Tropsch synthesis", Catalysis Today, 2002, 71.
4. Krishna, R., Van Baten, J.M., Urseanu, M.I. and Ellengerber, J., A scale up strategy for bubble column slurry reactors, *Catalysis Today*, **2001**, 66, pp. 199 - 207.

5. Dekwer, W.D., Access of Hydrodynamic Parameters Required in the Design and Scale-Up of Bubble Column Reactors, in *Chemical Reactors, ACS Symposium Series*, American Chemical Society: Washington, DC, 1981, pp. 213-241.
6. Liu, W., Mini-structured Catalyst Bed for Gas-Liquid-Solid Multiphase Catalytic Reaction, *AIChE J*, **2002**, July, pp. 1519-1532.
7. Post, M.F.M., Diffusion Limitation in Fischer-Tropsch Catalysts, *AIChE J*, **1989**, 35, 1007-1114.

## Chapter 8

# The Effect of Pressure on $CO_2$ Reforming of Methane and the Carbon Deposition Route Using Noble Metal Catalysts

**Abolghasem Shamsi and Christopher D. Johnson**

**National Energy Technology Laboratory, U.S. Department of Energy, 3610 Collins Ferry Road, Morgantown, WV 26505–0880**

Dry reforming of methane and $^{13}C$-labeled methane were studied over $Pt/ZrO_2$, $Pt/Ce\text{-}ZrO_2$, and $1\%wtRh/Al_2O_3$ at several pressures. It appears that carbon deposited on the catalysts resulted from both methane and $CO_2$. The presence of approximately equal ratios of $^{13}C$- and $^{12}C$-containing products suggests that the oxygen exchange reaction between CO and $CO_2$ is very fast. Temperature-programmed oxidation of carbon formed on the catalysts indicates that CO disproportionation contributes significantly to carbon deposition at lower and higher pressures. The Rh/alumina supported catalyst is the most resistant toward carbon deposition both at lower and higher pressures. In all catalysts, methane and $CO_2$ conversions as well as $H_2/CO$ ratios decreased with increasing pressure.

 

# Introduction

Natural gas, which is mostly methane, is primarily used for heating and electrical production. Currently, natural gas reserves are nearly as large as oil reserves and new reserves of natural gas are constantly being found. The economic use of natural gas, however, depends on its locality. It is not normally feasible to ship natural gas from remote areas to population centers. For instance, the north shore of Alaska has large natural gas reserves along with oil, but the natural gas is re-injected because the cost of transporting the gas is too expensive. Two available options at this time for bringing that resource to market are, building a pipeline, or cryogenic liquefaction. An alternative would be to chemically convert the gas to liquid and then transport it with the current infrastructure. To convert methane to liquid, it is first necessary to make synthesis gas and then convert the synthesis gas to liquid by methanol synthesis or Fischer-Tropsch reaction. The most costly part of the overall conversion is the production of synthesis gas.

Three different routes, or combinations of these, can be used to produce synthesis gas from methane. These are steam reforming, $CO_2$ reforming and partial oxidation. Each has its advantages and likewise has drawbacks. Steam reforming, which is a common industrial method of synthesis gas production, is very endothermic, as seen in the equations below. It also produces an $H_2/CO$ ratio of about 3/1, which is good for hydrogen production, but is too high for fuel synthesis. Methanol and Fischer-Tropsch synthesis use a $H_2/CO$ ratio of about 2/1.

$$CH_4 + CO_2 \leftrightarrow 2CO + 2H_2 \qquad \Delta H° = 247 \text{ kJ/mol}$$

$$CH_4 + H_2O \leftrightarrow CO + 3H_2 \qquad \Delta H° = 206 \text{ kJ/mol}$$

$$CH_4 + ½O_2 \leftrightarrow CO + 2H_2 \qquad \Delta H° = -35.7 \text{ kJ/mol}$$

$CO_2$ reforming of methane can be used to adjust the $H_2/CO$ ratio and provide the correct $H_2/CO$ ratio for Fischer-Tropsch synthesis and could potentially be used to reduce $CO_2$ emissions from other processes; however, it is even more endothermic than steam reforming. Partial oxidation is exothermic and has the correct $H_2/CO$ ratio for methanol synthesis, but requires a pure oxygen source, adding to the cost. In addition to the individual drawbacks, all of these processes must be run with O/C ratios of greater than 1 to prevent coking of the catalyst. This makes the processes more expensive in practice than would be expected under optimized conditions for the stoichiometric reactions. The propensity of these processes to form carbon at low O/C ratios is even more pronounced at high pressures where, in industry, they would need to be run to minimize reactor size and energy use. Clearly, creating new catalysts that kinetically inhibit carbon formation would make all of these reactions more efficient. We have chosen to focus on the $CO_2$ reforming of methane as a test reaction since carbon formation in this reaction is more pronounced, and because we are also interested in the potential of this reaction for $CO_2$ remediation.

Numerous research groups have focused their efforts on reducing carbon formation in the $CO_2$ reforming reaction.[1-8] For the most part, they have been carried out these experiments at atmospheric pressures, but several have also looked at carbon formation at high pressure. Tomishige et al.[9,10] have investigated the effects of pressure on dry reforming over a $Ni_{0.03}Mg_{0.97}O$ solid solution catalyst that, under atmospheric pressure, is very resistant to carbon deposition. Song et al. have reported preliminary results of high-pressure reaction over Ni/Na-Y and $Ni/Al_2O_3$ catalysts[11] as well as ruthenium catalysts[12] on the same supports. The effects of pressure are very similar in all cases, both methane and $CO_2$ conversion rates decreased, the $H_2/CO$ ratio decreased, and the rate of carbon deposition increased. Our own studies with $Pt/ZrO_2$, $Pt/Ce\text{-}ZrO_2$, and $Rh/Al_2O_3$ catalysts have given similar results.[13]

The higher rate of carbon deposition on reforming catalysts at high pressure leads naturally to the question of whether the source of carbon at elevated pressures is the same as at 1 atmosphere. It is known that the predominant sources of carbon on methane reforming catalysts are CO disproportionation and methane decomposition. An understanding which of these carbon formation routes is the primary contributor at high pressure is necessary information in developing catalysts that will prevent or reduce carbon formation under industrial reaction conditions. For this reason we have undertaken a $^{13}CH_4$ labeling study to compare the relative contributions of methane and $CO_2$ as the sources of carbon formation at low and high pressures.

## Experimental Section

The $Pt/ZrO_2$ and Ce-promoted catalysts were prepared as described by Stagg and Resasco.[14] All catalysts were reduced at 500°C for one hour before the reaction. The dry reforming reaction was performed in a fixed bed reactor. The reactor was a 0.3-m long ceramic reactor tube (6.35-mm o.d., 4.0-mm i.d.), sealed inside a stainless steel tube at higher pressures and contained a quartz thermocouple well. The reactor was loaded with 0.012 grams of catalyst mixed with 0.030 grams silica and held in place by quartz wool. The reactions were carried out at 800°C, and at 1 and 14 bar. Reaction products were monitored by a quadrupole mass spectrometer.

Temperature-programmed oxidation (TPO) of deposited carbon after the reaction was performed by heating the sample from 50 to 950°C at a rate of 20°C /min in flow of 2% oxygen in helium (20 $cm^3$/min). The TPO exit gases were also monitored by a quadrupole mass spectrometer.

## Results and Discussion

To investigate the source of carbon formed during dry reforming reactions on noble metal catalysts at high pressures, a mixture of $^{13}CH_4$ and $^{12}CO_2$ at an approximately 1:1 ratio was used. The reactions were carried out at both 1 and 14 atmospheres. Figure 1 shows the carbon oxide species formed, as monitored by quadrupole mass spectrometry, during the dry reforming reaction over a $Pt/ZrO_2$ catalyst for the 1 atm test. The flow rate of the reaction was 20 $cm^3$/min and temperature was 800°C. Note that for both CO and $CO_2$ the amount of these species containing $^{13}C$ is nearly equal to those containing $^{12}C$, with the $^{13}C$ compounds being slightly more prevalent. The presents of $^{13}CO_2$ in the reaction products suggest considerable reactions and back reactions of CO to $CO_2$ by either CO disproportionation, water gas shift, or oxygen exchange reactions. A similar test was done using $^{12}CH_4$, while monitoring for the same reaction products. It was found that the $^{13}C$-containing species were far less abundant. However, the ratio of $^{13}C$ to $^{12}C$ species was higher than would be expected from the naturally occurring isotopes, suggesting that there is a small amount of residual $^{13}C$ in the reaction system.

The same reaction of $^{13}CH_4$ with $^{12}CO_2$ over $Pt/ZrO_2$, at 800°C, was also carried out at 14 atm pressure. The carbon oxide species formed during the reaction were again monitored by quadrupole mass spectrometry. In this case, it was found that there a slightly higher amount of $^{12}C$ species relative to the $^{13}C$ species, as can be seen in Figure 2. The reason for the difference in relative amounts of the $^{13}C$ and $^{12}C$ containing species is not understood.

Identical experiments, to those carried out using the $Pt/Ce\text{-}ZrO_2$ catalyst, were also done using Ce-promoted catalyst, $Pt/Ce\text{-}ZrO_2$. The results of the tests performed on the $Pt/Ce\text{-}ZrO_2$ were similar to the results obtained when $Pt/ZrO_2$ catalyst was used. Monitoring of the reaction products using quadrupole mass spectroscopy showed approximately equal amounts of $^{13}C$ and $^{12}C$ in both the CO and $CO_2$ products for $^{13}CH_4$ as a reactant.

As expected, the ratios of $^{13}C$ carbon oxides to $^{12}C$ carbon oxides are significantly higher when $^{13}CH_4$ used in feed, see Figure 3. This result is for the reaction done over the $Pt/Ce\text{-}ZrO_2$ catalyst, but similar results were obtained for all catalysts. The presence of a significant amount of $^{13}CO_2$ supports the conclusion that oxygen exchange between $^{13}CO$ and $^{12}CO_2$ is very fast. However,

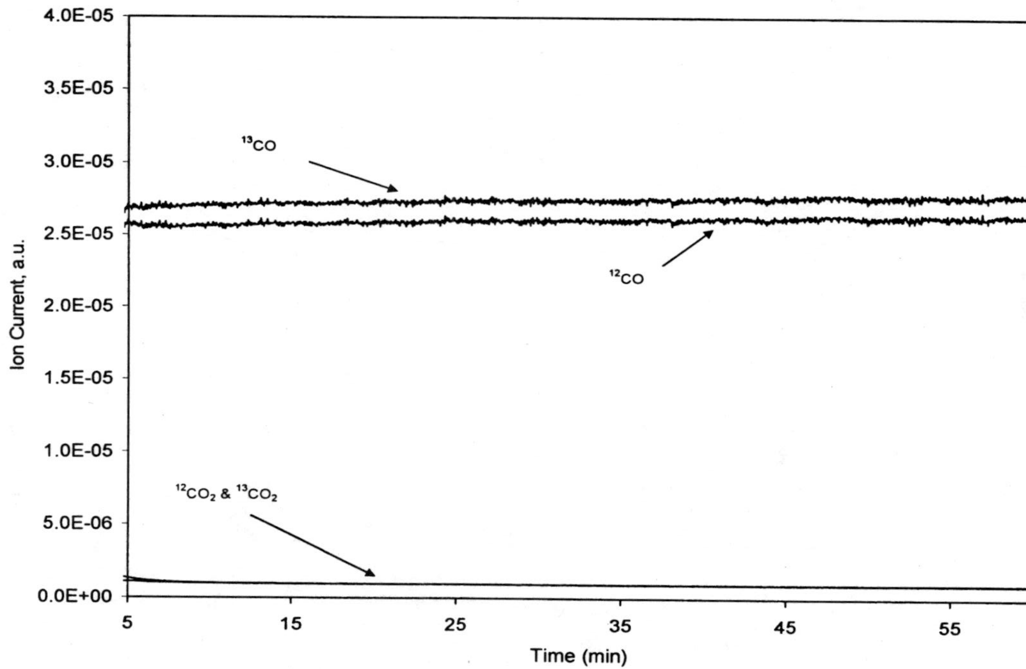


*Figure 1.. Product distribution over Pt/ZrO$_2$ at 800°C, $^{13}CH_4/CO_2$ = 1:1, 20cc/min, 1 atm.*

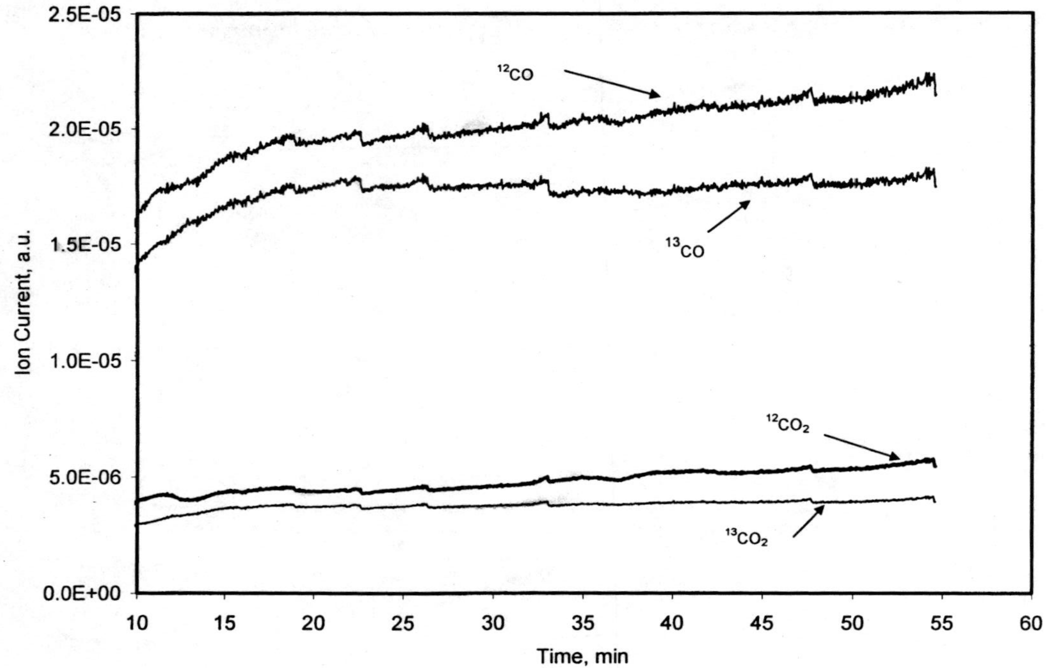


*Figure 2. Product distribution over $Pt/ZrO_2$ at 800°C, $^{13}CH_4/CO_2$ = 1:1, 20cc/min, 14 atm.*

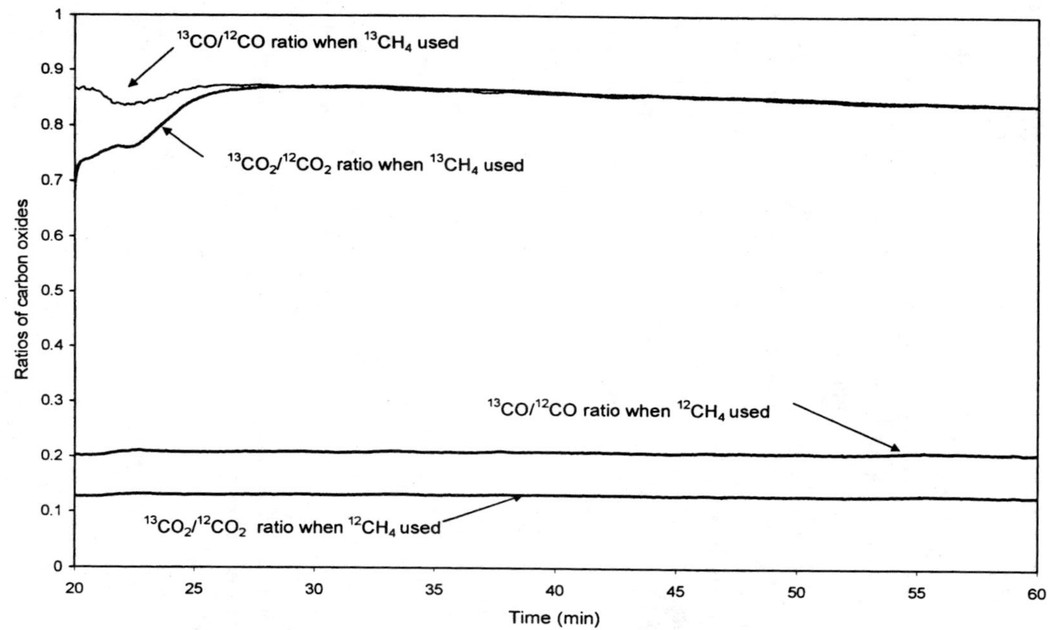


*Figure 3. Ratios of $^{13}CO$ and $^{13}CO_2$ in the reaction products when $^{12}CH_4$ and $^{13}CH_4$ reacted with $^{12}CO_2$ separately over Pt/Ce-$ZrO_2$ at 800°C, 1:12 feed ratio and 14 atm*

the ratios of $^{13}C$ and $^{12}C$ carbon oxide species detected when $^{12}CH_4$ was used as reactant were higher than would be expected given the natural occurrence of $^{13}C$.

Reactions of $^{12}CH_4$ and $^{12}CO_2$ over 1wt%/Rh/alumina were studied at 800°C and 1, 8, and 14 atm and the results are depicted in Figures 4-6, respectively. As shown in Figure 4 no significant deactivation of the catalyst was observed at atmospheric pressure during 5 hours of testing. However, as the pressure increased to 8 and 14 atm both methane and $CO_2$ conversion as well as $H_2/CO$ ratios decreased. The average conversions over 5 hours of reaction time decreased from 45 and 66% for methane and $CO_2$ respectively to 26 and 41% as the pressure increase from one to 14 atm. Furthermore, only a small amount carbon was detected in the catalyst bed at lower pressure but as the pressure increased the amount of carbon that formed increased, as indicated by the amount of $CO_2$ produced during temperature-programmed oxidation after 5 hours of reaction. Carbon deposition on Rh catalyst was very low in general, and so unlike previous experiments higher methane to $CO_2$ ratios was used in order to accelerate carbon formation.

Temperature-programmed oxidation (TPO) of the carbon formed (reactions performed at 800°C and pressures of one and 14 atm) on the $Pt/ZrO_2$, Ce-promoted ($Pt/Ce\text{-}ZrO_2$), and 1wt%Rh/Alumina catalysts were also performed. At lower pressure carbon deposition on these catalysts was very low and no significant amounts of $^{12}CO_2$ or $^{13}CO_2$ were produced during TPO. However, due to higher $CO_2$ conversion at lower pressure slightly higher amount of carbon containing $^{12}C$ was produced.

The results of the TPO experiment obtained for the Ce-promoted catalyst, reacted at 14 atm, are depicted in Figure 7. The TPO of the carbon formed on the $Pt/ZrO_2$ (Figure8) also showed the same peaks but at lower temperatures, around 480°C, for both $^{13}CO_2$ and $^{12}CO_2$. These results indicate that at high pressure these catalysts react similarly with respect to sources of carbon formation. Note that the ion current signals for $^{13}C$ and $^{12}C$ are essentially identical. This indicates that, at the pressure used for this reaction, both $^{13}CH_4$ and $CO_2$ contributed equally to carbon deposition. If the carbon was due to cracking of methane, then the majority of the deposited carbon would be $^{13}C$. This result is consistent with CO disproportionation as being the major contributor to carbon formation for $CO_2$ reforming reactions using noble metals at elevated pressures.

The similarity with respect to carbon formation at high pressure of these two catalysts is interesting in that it differs qualitatively from the carbon formation properties of these catalysts reported in an earlier study. Indeed, Stagg and Resasco[14] reported improved resistance to carbon deposition for the Ce-promoted catalyst for reactions done at atmospheric pressures. While the experiments reported here are not quantitative, the similarities in the results obtained for the two catalysts suggests that at high pressure the carbon formation

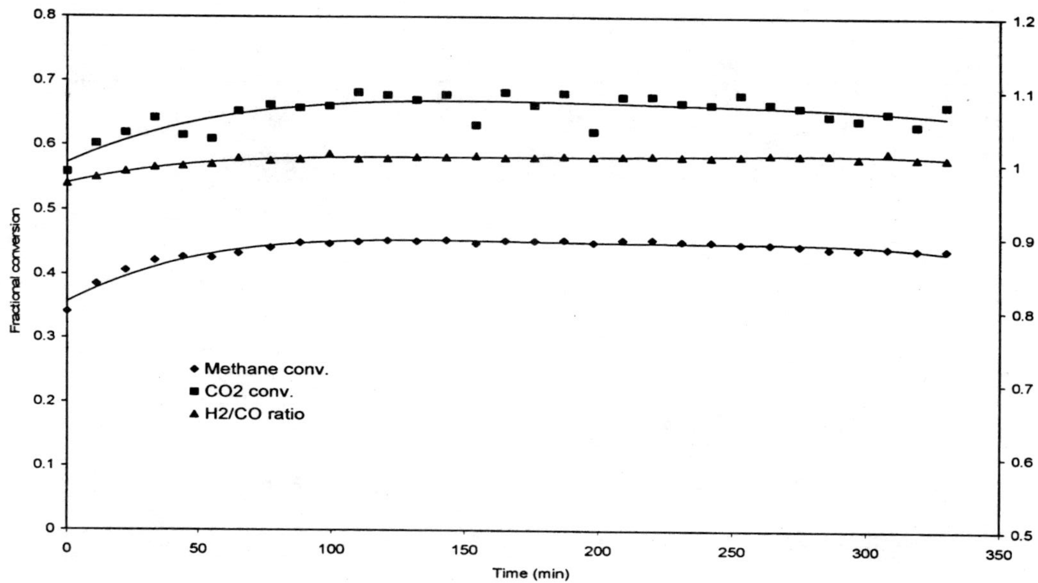


*Figure 4. Methane and $CO_2$ conversions over 1% Rh/Alumina at 800°C, $CH_4/CO_2$ = 2, and 1 atm.*

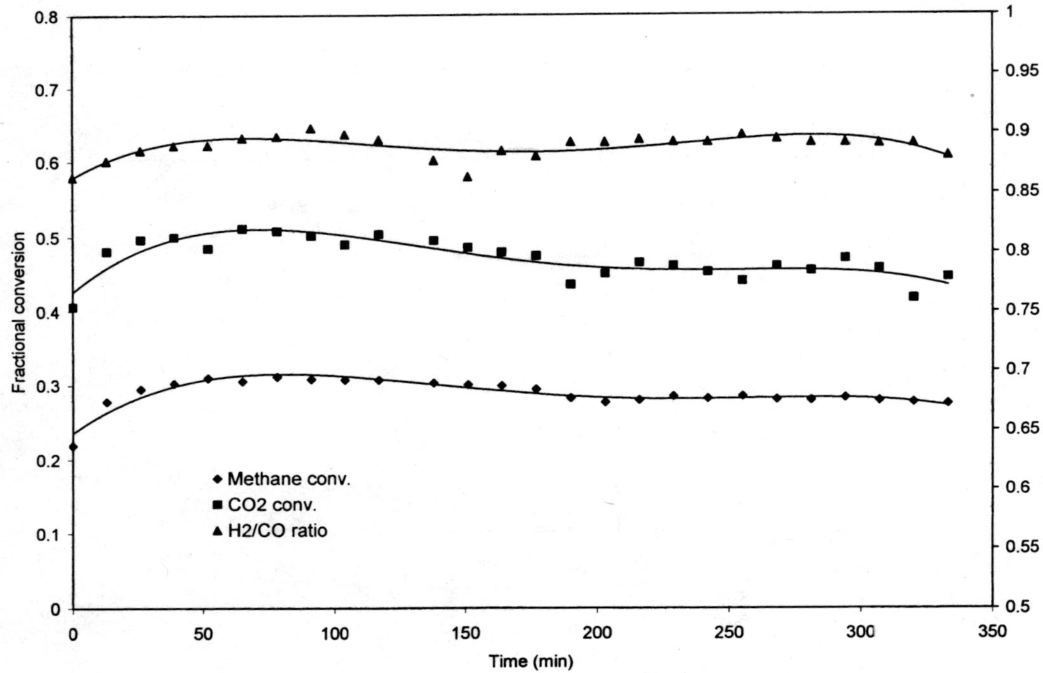


*Figure 5. Methane and $CO_2$ conversions over 1% Rh/Alumina at 800°C, $CH_4/CO_2$ = 2, and 8atm.*

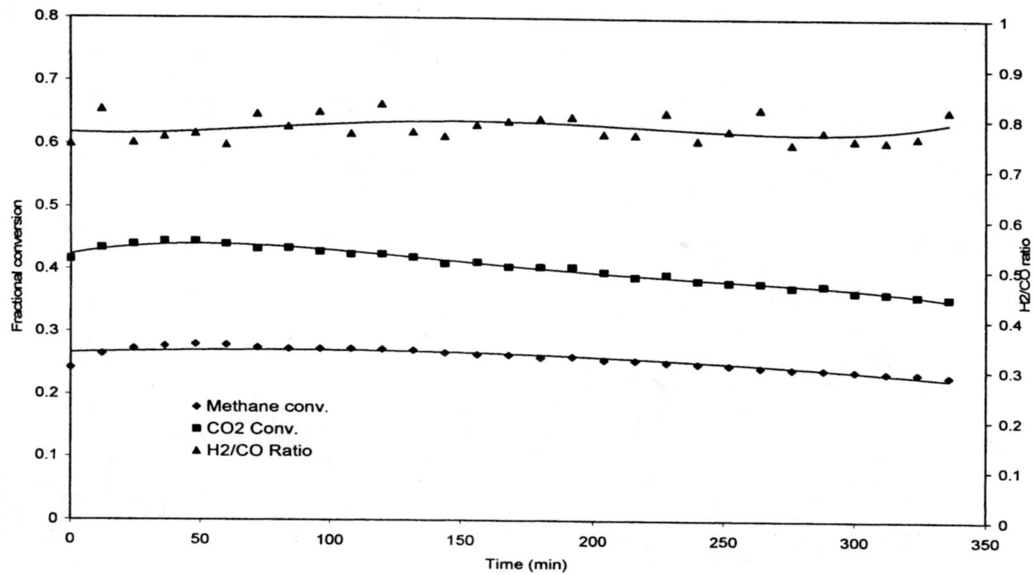


*Figure 6. Methane and $CO_2$ conversions over 1% Rh/Alumina at 800°C, $CH_4/CO_2$ = 2, 14 atm.*

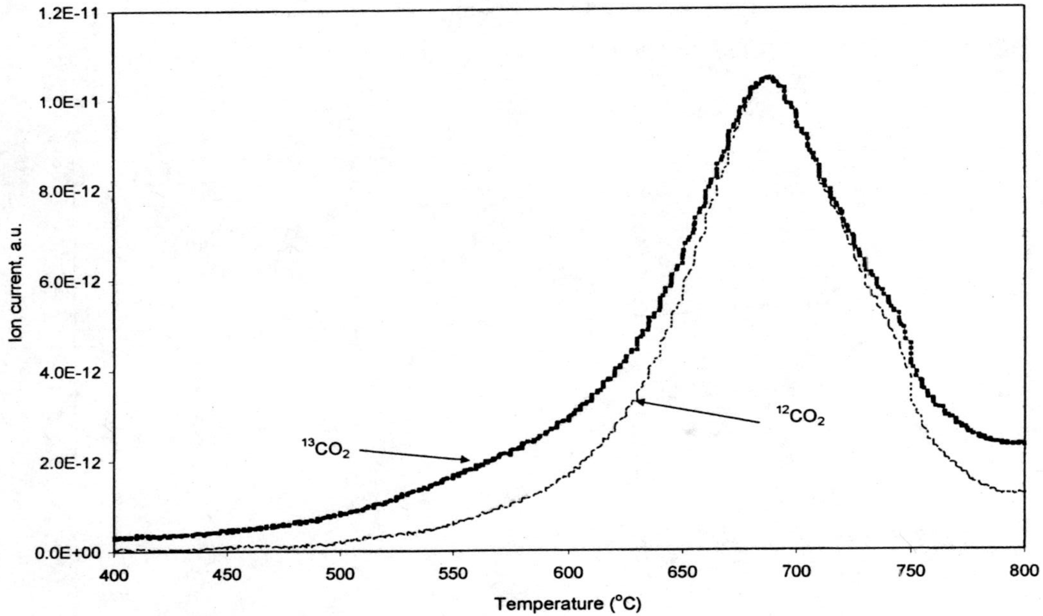


*Figure 7. Temperature-programmed oxidation of carbon deposited on Pt/Ce-ZrO$_2$ after reaction at 800°C, $^{12}CO_2$/$^{13}CH_4$ =1.2 feed ration, 14 atm*

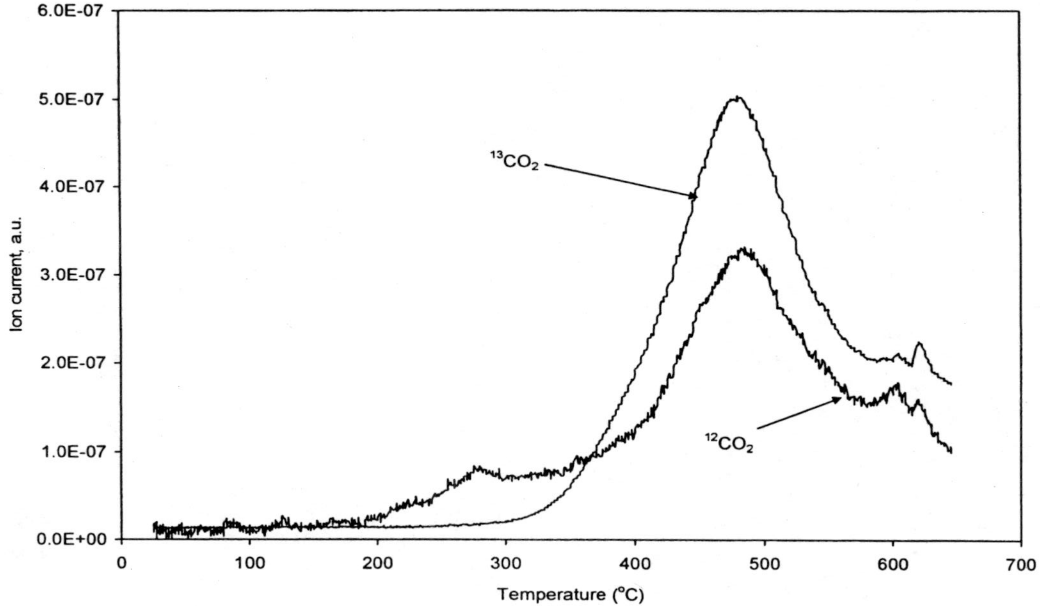


*Figure 8. Temperature-programmed oxidation of carbon deposited on Pt/$ZrO_2$ after reaction at 800°C, $^{12}CO_2$/$^{13}CH_4$ =1.2 feed ration, 14 atm*

route is the same, and is most likely due to CO disproportionation. This is consistent with thermodynamic calculations of $CO_2$ reforming of methane at elevated pressures, which show that as pressure is increased carbon formation due to this mechanism also increases.[13]

## Conclusion

Pressure has a significant effect on methane and $CO_2$ conversions for the dry reforming reaction, and influences the $H_2/CO$ ratios. Carbon formed on the catalysts comes from both methane and $CO_2$ as indicated by TPO experiments performed on carbon deposited during reactions conducted at 800°C (1 and 14 atm). As expected, the amount of carbon increases with increasing pressure. Since the exchange of oxygen between CO and $CO_2$ is very fast, almost identical amounts of carbon containing $^{12}C$ and $^{13}C$ were produced. These results are consistent with CO disproportionation being the predominant carbon deposition route at higher pressures.

## Acknowledgment

Financial support by the U.S. Department of Energy, Office of Fossil Energy, is gratefully acknowledged.

## References

1. Zhanlin Xu, Ming Zhen, Yingli Bi, Kaiji Zhe.n, Appl. Catal. A 198, 267-273 (2000)
2. Jung-Mei Wei, Bo-Qing Xu, Jin-Lu Li, Zhen-Xing Cheng, Qi-Ming Zhu, Appl. Catal. A 196, L167-L172 (2000)
3. Mitsunobo Ito, Tomohiko Tagawa, Shigeo Goto Appl. Catal. A 177, 15-23 (1999)
4. Shaobin Wang, G.Q. Lu, Energy and Fuels 12, 1235-1240 (1998)
5. S.-B. Tang, F.-L. Qiu, S.-J. Lu, Catal. Today 24, 253-255 (1995)
6. Ping Chen. Hong-Bin Zhang, Guo-Dong Lin, Khi-Rui Tsai, Appl. Catal. A, 166, 343-350 (1998)
7. J.D. Rostrup-Nielson, J.H. Bak-Hansen, J. Catal. 144, 38 (1993)
8. Jong-San Chang, Sang-Eon Park, Hakze Chon, Appl. Catal. A, 145, 111-124 (1996)
9. K. Tomishige, Y. Himeno, O. Yamazaki, Y Chen, T. Wakatsuki, K. Fujimoto, Kinetics and Catal. 40(3) 388-394 (1999)

10. K. Tomishige, Y. Himeno, Y. Matsuo, Y. Yoshinaga, K. Fujimoto, Ind. Eng. Chem. Res. 39, 1891-1897 (2000)
11. C. Song, S.T. Srinivas, L. Sun, J.N. Armor, Chem. Soc., Div. Petrol Chem., Preprints., 45 143-148 (2000)
12. S.T. Srinivas, C. Song, Am. Chem. Soc., Div. Petrol Chem., Preprints., 45 153-156 (2000)
13. A. Shamsi, C.D. Johnson, Am. Chem. Soc., Div. Fuel Chem., Preprints, 46, 94-96 (2001)
14. S.M. Stagg and D.E Rasasco, Stud. Surf. Sci. Catal, 1998, Vol. 119, 813-818.

# Advances in Catalysts

# Chapter 9

# Direct Decomposition of Methane to Hydrogen on Metal-Loaded Zeolite Catalyst

**Lucia M. Petkovic, Daniel M. Ginosar, Kyle C. Burch, and Harry W. Rollins**

**Chemistry Department, Idaho National Laboratory, P.O. Box 1625, MS 2208, Idaho Falls, ID 83415–2208**

The regeneration of a Ru-Mo/ZSM5 catalyst used to produce hydrogen by decomposition reactions was examined at 873 K in either flowing hydrogen or air. The Ru-Mo/ZSM5 catalyst was deactivated under methane decomposition reaction conditions at temperatures of 873, 973, and 1073 K. The effect of reaction and regeneration conditions on catalyst activity, BET surface area and temperature programmed oxidation profiles is discussed. Regeneration using hydrogen was found to be less effective than air oxidation for the conditions explored. The methane decomposition reactions produced hydrogen primarily through the formation of liquid aromatics. The highest hydrogen production rates were obtained at 1073 K, however, deactivation was greatest at this temperature and the catalyst was not fully regenerated. A better selectivity to liquids combined with a relatively stable catalyst activity indicates that 973 K may be adequate for longer term catalyst and cycled regeneration analyses.

The manufacture of hydrogen from natural gas is essential for the production of ultra clean transportation fuels. Not only is hydrogen necessary to upgrade low quality crude oils to high-quality, low sulfur ultra-clean transportation fuels, hydrogen could eventually replace gasoline and diesel as the ultra-clean transportation fuel of the future. Currently, refinery hydrogen is produced through the steam reforming of natural gas. Although efficient, the process is responsible for a significant portion of refinery $CO_2$ emissions. The steam reforming process is capital intensive, owing to the high-temperature, highly endothermic reaction, the need for water gas shift reactors, and gas-gas separation processes. Further, the hydrogen product is contaminated with CO at concentrations that are detrimental to proton exchange membrane (PEM) fuel cells.

Alternative direct methane conversion chemistries are available that do not produce $CO_x$ byproducts and offer the potential of lowering capital equipment costs. Methane can be decomposed to produce hydrogen with product formation of carbon, aromatics, or alkene gasses, as represented below.

$$CH_4 \rightarrow C + 2\,H_2 \qquad \Delta H = 75 \text{ kJ/mol}$$
$$CH_4 \rightarrow 1/6\,C_6H_6 + 3/2\,H_2 \qquad \Delta H = 89 \text{ kJ/mol}$$
$$CH_4 \rightarrow \tfrac{1}{2}\,C_2H_4 + H_2 \qquad \Delta H = 101 \text{ kJ/mol}$$

These reactions are about half as endothermic as the main steam reforming reaction ($\Delta H$ = 206 kJ/mol), and, since the carbon value is captured as salable products, $CO_2$ emissions are reduced by at least 80%. However, less hydrogen is produced per unit of natural gas feed, for both reaction and heat requirements. Thus, the value of the carbon based product has a significant impact on the economic viability of the reaction process.

The production of hydrogen from methane over zeolite supported metal catalysts can be examined as an alternative to steam reforming because the concomitant aromatization reactions can increase the economic potential of the process. For methane aromatization, Mo/ZSM5 catalysts have been intensively studied since their first report in 1993 (*1, 2*). In 1997 (*3*), the promotional effect of ruthenium over Mo/ZSM5 catalysts was reported. Other second metals have also been studied to improve catalyst activity and stability and a review on this topic is available (*4*).

Larachi et al. (*5*) and Iliuta et al. (*6, 7*) compared the conversion of methane over Ru-Mo/ZSM5 catalysts with and without the presence of a $H_2$ permeating membrane and concluded that the availability of $H_2$ helped in the hydrogenation of carbon deposits and hence restoration of catalyst active sites. For the work presented here, it was hypothesized that cycling treatments under flowing $H_2$ might also restore catalyst active sites. Oxidative regeneration under flowing air was also performed for comparison purposes.

## Experimental

A commercially available HZSM5 zeolite (Zeolyst CBV 3020E) was co-impregnated with aqueous solutions of ammonium heptamolybdate tetrahydrate (99.98%, Aldrich) and ruthenium (III) chloride hydrate (99.98%, Aldrich) by the incipient wetness technique to obtain a final metal content of 3% Mo and 0.5% Ru. The material was dried at 393 K for 2 h. Five hundred milligrams of the dried material were loaded in a tubular quartz reactor, calcined under flowing air (Zero grade, US Welding) for 4 h at 873 K, then for 0.5 h at 973 K, and reduced under flowing hydrogen for 0.5 h at 573 K. The temperature was increased to reaction temperature (i.e., 873, 973, or 1073 K) and a flow of 20 $cm^3$/min of pure methane (99.99%, Scott Specialty Gases, PA) was initiated. Hydrogen and liquid product yields, as a function of time on stream (TOS), were measured using an on-line Hewlett Packard 5890 gas chromatograph equipped with both FID and TCD detectors and Carboxen 1010 Plot and Petrocol columns. Off-line analysis was performed to quantify ethylene and propylene yields, and to confirm the absence of $CO_x$ contaminants.

Three sets of experiments were performed to examine the effects of regenerating catalysts with either air or hydrogen: the first set examined the effect of reaction on the catalyst; the second set was used to produce a catalyst sample that had been exposed to both reaction and regeneration; and the last set followed a full cycle of reaction/regeneration/reaction conditions. The first set of experiments was terminated at 3 h TOS and the spent catalyst recovered. For the second set of experiments, at 3 h TOS a regeneration step with either 30 $cm^3$/min flowing air or 20 $cm^3$/min flowing hydrogen was initiated. The temperature was set at 873 K and the gaseous regenerant was flowed through the catalyst bed for 1 hour.

The third set of experiments was performed to determine catalyst behavior after regeneration. For these experiments, after 1 h regeneration at 873 K, the temperature was set at reaction temperature, and a flow of 20 $cm^3$/min of pure methane (99.99%, Scott Specialty Gases, PA) was initiated again. For the experiments in which the regenerant was air, a 30 min reduction treatment under 30 $cm^3$/min flowing $H_2$ at 573 K was applied after regeneration. In all experiments where gas flow was switched from methane to regenerant, or back, the reactor was purged with flowing argon.

Spent catalyst samples were recovered and submitted to ex-situ characterization analyses following the experiments.

B.E.T. surface areas and micropore volumes were determined on a Quantachrome Autosorb 1-C apparatus at liquid nitrogen temperatures. B.E.T. surface areas were calculated in the range of P/Po between 0.05 and 0.10. Micropore volumes were as determined by the t-plot method.

The carbon material deposited on the catalyst was characterized by temperature-programmed oxidation (TPO) on a Perkin Elmer Diamond TG/DTA microbalance under 100 ml/min flowing air. About 10 mg of sample were placed in the balance pan and heated from room temperature to 373 K at 10 K/min. The temperature was kept at 373 K for 30 min, increased to 1073 K at 10 K/min, and then decreased to 373 K. The change in weight between the two temperature segments at 373 K was utilized to report carbon content. The negative of the derivative of weight with respect to time was utilized to obtain the TPO profiles.

Transmission electron microscopy analyses were performed on a Philips EM 420 at 120kV. A small amount of catalyst was dispersed in ethanol and a drop of this dispersion, dried on a carbon-coated copper grid at room temperature, was utilized for the analysis.

Diffuse reflectance infrared Fourier transform spectroscopy (DRIFTS) studies were performed at 373 K on a Nicolet Magna 750 Fourier transform infrared spectrometer equipped with a commercial Spectratech diffuse reflectance cell. Eight-hundred-scan spectra were collected in the 4000-400 $cm^{-1}$ range at a resolution of 4 $cm^{-1}$. The catalyst samples were diluted in KBr at a ratio of about 1:200 sample:KBr for DRIFTS analyses.

## Results and Discussion

### Reaction studies

Hydrogen and benzene concentrations in the product stream as a function of time on stream (TOS) at different reaction temperatures are shown in Figure 1. Minor amounts (not shown) of toluene, naphthalene, ethylene, propylene, and other aromatics were also found in the product stream. Although formation of carbon oxides at short TOS upon contacting Mo/ZSM5 catalysts with methane has been reported in the literature (*8*), these products were not detected in any of the gas samples analyzed. Formation of carbon oxides is reported to occur during an induction period in which catalytic active sites are developed by carburization of Mo species. An inverse relation between methane hourly space velocity (MHSP) and induction time has been reported (*6*). Hence, the absence of carbon oxides even on the first chromatographic sample may be assigned to an induction period shorter than 20 min due to the relatively larger (i.e., 2400 mL/(h g)) MHSP applied here.

As expected from thermodynamic and kinetic considerations, higher initial conversions were seen at higher reaction temperatures. The concentration of $H_2$ was about 5% at 873 K and 15% at 973 K when fresh catalyst samples were submitted to reaction conditions. Although some decrease with TOS in the production of $H_2$ was seen at those two temperatures, a more pronounced decrease was seen at 1073 K, where $H_2$ concentration changed from about 27% to 12%.

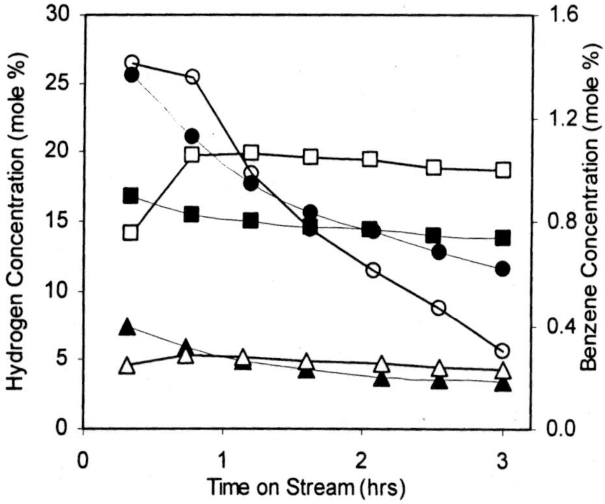


*Figure 1. Effect of TOS on hydrogen (filled symbols) and benzene (hollow symbols) concentration in the product stream at 873 (triangles), 973 (squares), and 1073 K (circles).*

The catalyst samples recovered after reaction at 873, 973, and 1073 K contained 3.1, 4.5, and 11.1 wt% of carbonaceous deposits, respectively. Assuming a $C_1H_1$ stoichiometry for carbonaceous deposits and adopting average concentrations of aromatics and olefins (i.e., ethylene and propylene), an approximate mass balance was calculated to estimate the contribution of the different reactions to the production of hydrogen. As seen on Table I, at the three temperatures studied, the reactions producing liquid aromatic products accounted for most of the hydrogen produced. It is worth noting that the three chosen reactions are not independent. Aromatics such as benzene are produced by polymerization of ethylene. Further aromatization and dehydrogenation of benzene and higher aromatics contributes to carbonaceous deposits.

## Regeneration studies

After treating the spent catalyst samples with flowing air (see Figures 2 and 3), the profiles of $H_2$ and benzene resembled the ones obtained from fresh samples at 873 and 973 K. However, at 1073 K some loss of activity was apparent.

**Table I. Percent contribution of different reactions to $H_2$ production**

| *Reaction temp. (K)* | *Benzene* | *Total liquid aromatics* | *Ethylene* | *Carbonaceous deposits* |
|---|---|---|---|---|
| 873 | 49 | 72 | 0.8 | 27 |
| 973 | 61 | 87 | 0.7 | 12 |
| 1073 | 54 | 67 | 2.8 | 30 |

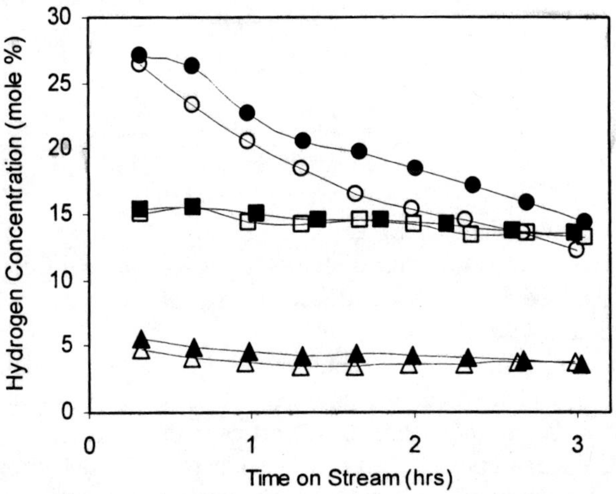


*Figure 2. Effect of regeneration with air on hydrogen yields at 873 (triangles), 973 (squares), and 1073 K (circles). Filled and hollow symbols correspond to reaction on fresh and regenerated samples, respectively.*

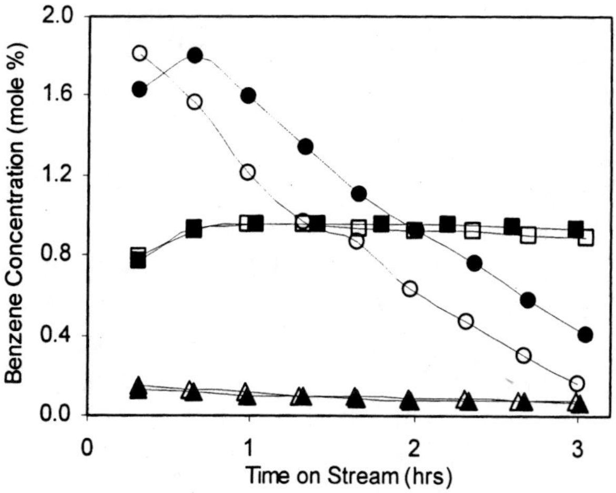


*Figure 3. Effect of regeneration with air on benzene yields at 873 (triangles), 973 (squares), and 1073 K (circles). Filled and hollow symbols correspond to reaction on fresh and regenerated samples, respectively.*

Submitting spent catalyst samples to hydrogen treatment was less effective than air treatment as seen in Figures 4 and 5. Working at reaction temperatures of 973 K or lower, Larachi et al. (*5*) found that non-carbidic coke formation was reversible and active sites were restored when $H_2$ was available either during off-membrane operation or $CH_4/H_2$ regenerative steps. The results presented here were obtained at higher MHSP and show that the catalyst recovered some activity only at temperatures of 873 and 973 K. However, at the highest temperature of 1073 K, the hydrogen regeneration was not effective, and profiles of $H_2$ and benzene production obtained during the second reaction step (hollow symbols in Figures 4 and 5) seemed to continue the same deactivation trend shown during the first reaction step (filled symbols in Figures 4 and 5).

Table II shows the amounts of carbon deposits as determined by TPO analyses. As reaction temperature increased, the amount of carbon deposits retained on the catalysts increased. Regenerating the spent catalyst with air produced an almost complete elimination of carbon residues. In contrast, regeneration with $H_2$ produced only a partial elimination of deposits. TPO profiles associated with the carbon deposits are shown in Figures 6-8.

Figure 6 provides a comparison of TPO profiles before and after regeneration in air and hydrogen. TPO profiles for samples that were deactivated at lower reaction temperatures in the first reaction step were shifted to the left (thick lines in Figure 6) indicating that the carbon deposits were easier to oxidize

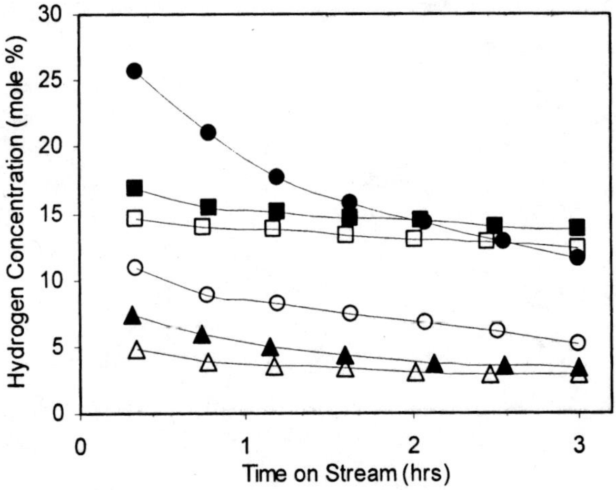


*Figure 4. Effect of regeneration with hydrogen on hydrogen yields at 873 (triangles), 973 (squares), and 1073 K (circles). Filled and hollow symbols correspond to reaction on fresh and regenerated samples, respectively.*

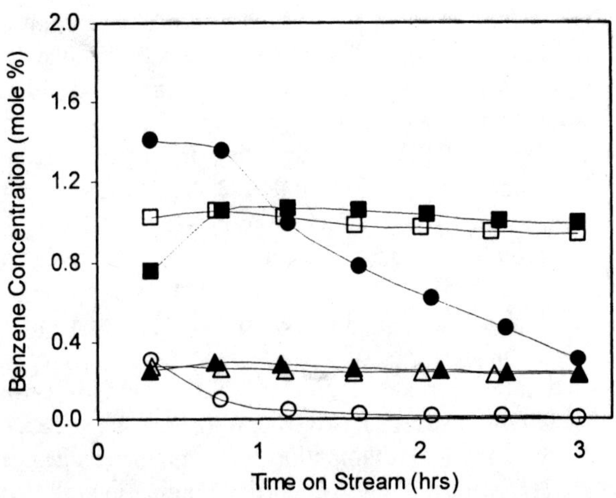


*Figure 5. Effect of regeneration with hydrogen on benzene yields at 873 (triangles), 973 (squares), and 1073 K (circles). Filled and hollow symbols correspond to reaction on fresh and regenerated samples, respectively.*

**Table II. Carbon deposits (wt%) as determined by TPO**

| *Reaction temp. (K)* | *1st reaction step* | *Regener. with air* | *Regener. with $H_2$* | *2nd reaction after air* | *2nd reaction after $H_2$* |
|---|---|---|---|---|---|
| 873 | 3.1 | 0.6 | 2.8 | 2.0 | 3.6 |
| 973 | 4.5 | 0.0 | 3.8 | 4.6 | 4.8 |
| 1073 | 11.1 | 0.7 | 5.8 | 9.7 | 13.7 |

during the analyses. Although this may be ascribed to less condensed hydrocarbons, the lower carbon content may also decrease diffusion limitations that retard the exiting of oxidation products from the zeolitic catalyst.

The TPO profiles of samples that after regeneration were submitted again to reaction conditions are shown in Figures 7 and 8. When the regenerating treatment was under flowing air (Figure 7), the carbon residues retained during the second time under reaction conditions (thin lines) showed similar TPO profiles as the ones obtained after the first time under reaction conditions (thick lines). However, regenerating the catalyst with $H_2$ seemed to have altered the

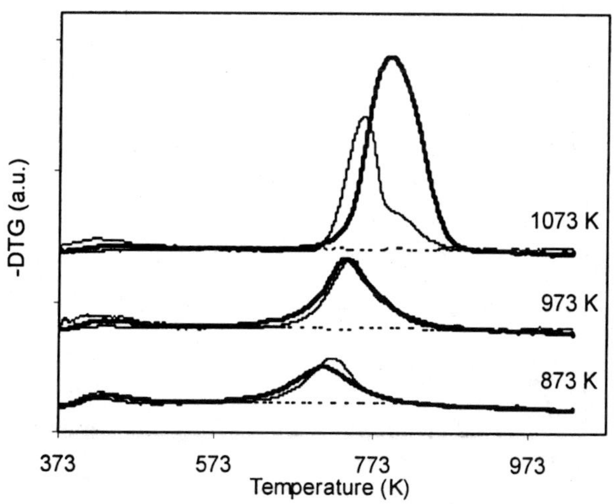


*Figure 6. Comparison of TPO profiles before (thick lines) and after regeneration in air (dotted line) and $H_2$ (thin line). Reaction temperature as indicated.*

nature of the residues particularly when the amounts were high. The material still left on the catalyst after $H_2$ regeneration plus the newly retained material during the second reaction at 1073 K produced a two-peak TPO profile as shown in Figure 8. This indicates that although the regeneration treatment utilizing $H_2$ was not absolutely effective in freeing the catalyst from carbonaceous deposits, it altered the nature of the deposits.

DRIFTS analysis was performed on the spent samples to obtain information on the chemical nature of the carbonaceous deposits retained on the catalysts. The spectra of all samples submitted to reaction and the spectra of samples regenerated under flowing $H_2$ showed a band at 1580 $cm^{-1}$, which is characteristic of highly condensed hydrocarbons and graphitic-like species. This band was absent on the spectra of samples regenerated under flowing air. C-H stretching bands were hardly noticeable on any spectra which confirm the condensed nature of the deposits.

Transmission electron microscopy analyses showed a homogenous covering of the zeolite surface with carbon deposits. The highly condensed and likely graphitic-type nature of the carbon deposits agrees with the results presented by Larachi et al. (*5*). These authors identified three carbon-containing species on spent Ru-Mo/ZSM5 by X-ray photoelectron spectroscopy (XPS) analyses. The most prominent peak was assigned to graphitic-like carbon, a weaker peak to aromatic-aliphatic species, and a hardly visible peak to carbidic carbon. As

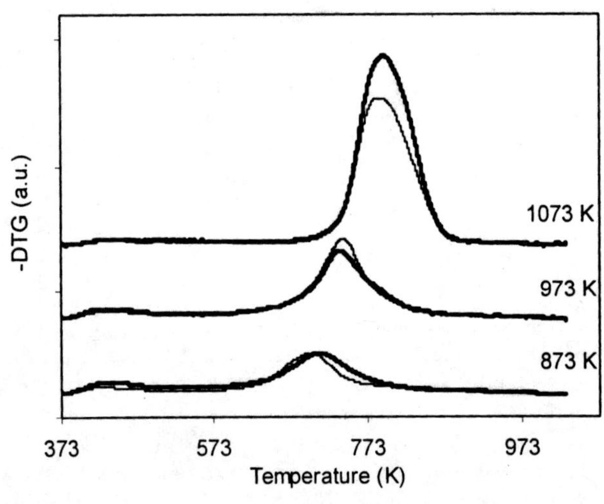


*Figure 7. Effect of regeneration with air on TPO profiles. Thick and thin lines correspond to samples submitted once and twice to reaction conditions, respectively. Reaction temperatures as indicated.*

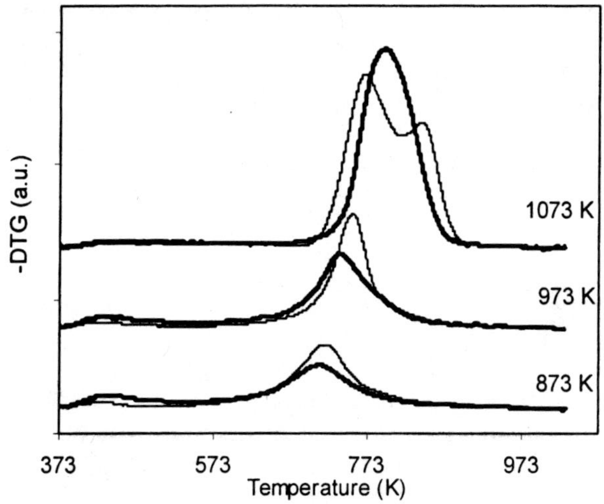


*Figure 8. Effect of regeneration with $H_2$ on TPO profiles. Thick and thin lines correspond to samples submitted once or twice to reaction conditions, respectively. Reaction temperatures as indicated.*

shown by Ding et al. (*9*), the carbidic carbon is observed on TPO profiles around 500 K when samples are contacted with methane for only short TOS. After longer TOS only one TPO peak is apparent in agreement with the results presented here.

Table III shows the results of surface area and micropore volume analyses. For the first reaction step higher carbon contents (Table II) were associated with higher losses of surface area and micropore volume, as expected. The fresh sample displayed lower surface area and pore volume than the one utilized at 873 K. This may be due to experimental variability and/or to an actual change in surface area due to changes in the nature of the molybdenum-containing species from oxide to carbide. The highly condensed structure (i.e., likely graphitic-type) of the carbon deposits agrees with the measured micropore volumes when considering that graphite density is around 2 g/cm$^3$. For example, a fresh zeolite displaying a micropore volume of 0.12 cm$^3$/g would lose about 0.06 cm$^3$/g if graphitic carbon deposits reached 11.1%.

Air regeneration produced an important recovery of micropore volume. Even though the amount of carbon deposited during the second step under reaction conditions was about the same or lower, surface areas and micropore volumes were lower than those at the end of the first step. Although some pore plugging might explain these results, it is more likely that irreversible changes in the zeolite structure during air regeneration lead to lower micropore volumes.

**Table III. Catalyst B.E.T. surface areas and micropore volumes**

| *Reaction temp. (K)* | *1st reaction step* | *Regener. with air* | *Regener. with $H_2$* | *2nd reaction after air* | *2nd reaction after $H_2$* |
|---|---|---|---|---|---|
| 873 | 407 (0.13) | 392 (0.12) | 375 (0.12) | 307 (0.09) | 316 (0.10) |
| 973 | 310 (0.10) | 385 (0.12) | 336 (0.11) | 307 (0.09) | 222 (0.08) |
| 1073 | 216 (0.07) | 375 (0.12) | 174 (0.04) | 190 (0.06) | 46 (0.006) |

NOTE: Surfaces areas in $m^2/g$; micropore volumes in parenthesis, $cm^3/g$. Fresh sample: 403 $m^2/g$, 0.12 $cm^3/g$.

Rapid burning of carbon deposits by direct introduction of air at such high temperatures has been reported to result in destruction of the ZSM5 crystal structure and complete loss of activity for methane aromatization after succesive air regenerations (*10*). On the other hand, $H_2$ regeneration was less effective. A decrease in micropore volume was even measured after hydrogen regeneration on the sample initially utilized at 1073 K. This decrease in micropore volume (from 0.07 to 0.04 $cm^3/g$) along with the decrease in carbon content (from 11.1 to 5.8%) indicates that the carbon deposits became bulkier/more hydrogenated and/or some pore plugging occurred during regeneration. Pore plugging likely increased during the second step under reaction conditions leading to an almost complete loss of micropore volume by the end of the experiment at 1073 K.

## Conclusions

Although hydrogen production rates were higher at 1073 K, the rapid decrease in catalyst activity, the incomplete activity recovery by either regenerant, and the higher selectivity to carbon deposits makes 1073 K unfavorable for continued or cycled application of this catalyst. A better selectivity to liquids combined with a relatively stable catalyst activity indicates that 973 K may be adequate for longer term catalyst and cycled regeneration analyses. It is expected that cycled $H_2$ regeneration will be milder with respect to zeolite structure stability when compared with air regeneration but longer term stability and meticulous economic analysis of this process are necessary before any application of this catalyst for production of hydrogen is undertaken.

## Acknowledgements

This work was supported by the U.S. Department of Energy Office of Fossil Energy under DOE Idaho Operations Office Contract DE AC07 05ID14517.

## References


1. Wang, L.; Tao, L.; Xie, M.; Xu, G.; Huang, J.; Xu, Y. *Catal. Lett.* **1993**, *21,* 35-41.
2. Lacheen, H. S.; Iglesia, E. *J. Catal.* **2005**, *230,* 173-185.
3. Shu, Y. Y.; Xu, Y. D.; Wong, S. T.; Wang, L. S.; Guo, X. X. *J. Catal.* **1997**, *170,* 11-19.
4. Xu, Y.; Lin, L. *Appl. Catal. A* **1999**, *188,* 53-67.
5. Larachi, F.; Oudghiri-Hassani, H.; Iliuta, M. C.; Grandjean, B. P. A.; McBreen, P. H. *Catal. Lett.* **2002**, *84,* 183-192.
6. Iliuta, M. C.; Larachi, F.; Grandjean, B. P. A.; Iliuta, I. *Ind. Eng. Chem. Res.* **2002**, *41,* 2371-2378.
7. Iliuta, M. C.; Grandjean, B. P. A.; Karachi, F. *Ind. Eng. Chem. Res.* **2003,** *42,* 323-330.
8. Jiang, H.; Wang, L.; Cui, W.; Xu, Y. *Catal. Lett.* **1999**, *57,* 95-102.
9. Ding, W.; Li, S.; Meitzner, G. D.; Iglesia, E. *J. Phys. Chem. B* **2001**, *105,* 506-513.
10. Borry, R. W.; Iglesia, E. Proceedings of the Natural Gas Conference, Emerging Technologies for the Natural Gas Industry, P14, March 24 to 27, 1997, Houston, Texas. Available on-line at http://www.netl.doe.gov/publications/proceedings/97/97ng/ng97_pdf/NGP14.PDF on 12-01-2005.

## Chapter 10

# Novel Catalytic Properties of Pd Sulfide for the Synthesis of Methanol from Synthesis Gas in the Presence and Absence of $H_2S$

## Effect of Several Additives on the Methanol Synthesis Activity of Pd Sulfides

**Naoto Koizumi, Kazuhito Murai, Satoshi Takasaki, and Muneyoshi Yamada***

**Department of Applied Chemistry, Graduate School of Engineering, Tohoku University, Aoba 6–6–07, Aramaki, Aoba-ku, Sendai 980–8579, Japan**

Catalytic activity of palladium (Pd) sulfides for the synthesis of methanol from the syngas (a mixture of carbon monoxide (CO) and hydrogen ($H_2$)) was investigated in the present study in the presence and/or absence of $H_2S$. Pd sulfide supported on $SiO_2$ was modified with several additives, and their methanol synthesis activity was investigated at various reaction conditions. Modification of the Pd sulfide supported on $SiO_2$ with alkali metal and/or alkaline earth metal additives improved the methanol synthesis activity greatly. The maximum space-time yield of methanol was obtained with the sulfided Ca/Pd/$SiO_2$ (730 g kg-cat$^{-1}$ h$^{-1}$) at 593 K and 5.1 MPa in the absence of $H_2S$, which was much higher than that with the ZnO/$Cr_2O_3$ catalyst, and was about 50% of that with the commercial Cu/ZnO/$Al_2O_3$ catalyst (from CO/$H_2$/$CO_2$ feed) at the same reaction conditions. Although both the ZnO/$Cr_2O_3$ and the Cu/ZnO/$Al_2O_3$ catalysts were seriously deactivated in the presence of $H_2S$ (100 ppm in concentration), this supported sulfide preserved a constant space-time yield of methanol (250 g kg-cat$^{-1}$ h$^{-1}$) even in the presence of $H_2S$. This space-time yield of methanol was comparable with that reported for the K/$MoS_2$ catalyst in the presence of $H_2S$ at much higher reaction pressure (5 MPa vs. 21 MPa). The role of the additives was also discussed based on the results obtained by several spectroscopic techniques, such as *in-situ* diffuse reflectance FTIR spectroscopy.

## Introduction

Reduced Pd catalysts are well known to be active and selective for the synthesis of methanol from syngas (CO/$H_2$) in the presence of some metal oxides, such as alkali metal oxides (*1,2*), alkaline earth metal oxides (*2,3*), and/or lanthanide oxides (*2,4-6*). The selectivity for methanol on the reduced Pd catalysts in the presence of these metal oxides is typically above 90-mol%, while the metallic Pd particles supported on $SiO_2$ containing impurities in sub-ppm levels are selective for the synthesis of hydrocarbon rather than methanol. Recent study has shown that the reduced Pd/$CeO_2$ yields 310 g kg-cat$^{-1}$ h$^{-1}$ of methanol at 473 K and 2.0 MPa (*6*). The methanol synthesis with a commercial Cu/Zn/Al catalyst, on the other hand, requires a higher reaction temperature (523 K) to obtain a comparable methanol production (*7*). Thus, the reduced Pd catalyst is a most promising candidate for the low-temperature methanol synthesis catalyst. Concerning the role of metal oxides, different speculations have appeared in the literatures. Ponec *et al.* (*3,8*) proposed that the promoter oxides create Pd cation sites, where formyl species is formed from CO and $H_2$ molecules. The formyl species is regarded as the crucial intermediate for the synthesis of methanol (*9*). On the other hand, Tamaru *et al.* (*1*) observed that the amount of formate species formed by the adsorption of formaldehyde on reduced Pd/$SiO_2$ promoted with Na oxide correlates with the formation rate of methanol. Isotopic experiments suggested that oxygen atoms, which are available for the formation of the formate species from CO and $H_2$ molecules, are preserved by the promoter oxide.

Recent study found that Pd sulfide prepared by sulfiding $PdCl_2$ (structure determined with XRD was $Pd_{16}S_7$) yielded methanol with a selectivity above 95-mol% (*10*). This result is notable because, so far, only Mo and W sulfides have been known to show weak activities for the synthesis of gaseous hydrocarbons from syngas (*11*). This type of sulfides had some attentions because Mo sulfide modified with alkali metal additives showed a constant activity for alcohol synthesis even in the presence of $H_2S$ (*11*). This is in marked contrast with the

$Cu/ZnO/Al_2O_3$ and the reduced Pd catalysts that are seriously deactivated in the presence of $H_2S$ (*12,13*). In the case of $Pd_{16}S_7$, it was found to preserve a constant methanol yield even in the presence of $H_2S$ 100 ppm in concentration. Thus, Pd sulfide is a new type of sulfur tolerant catalysts. However, the space-time yield of methanol obtained with $Pd_{16}S_7$ was 80 g kg- $cat^{-1}$ $h^{-1}$ at 613 K and 5.1 MPa in the absence of $H_2S$, which was no more than 10% of that obtained with a commercial $Cu/ZnO/Al_2O_3$ catalyst from $CO/H_2/CO_2$ feed. Then, we investigated the activity and selectivity of Pd sulfides supported on several metal oxides for the CO hydrogenation. Among the metal oxide ($M_xO_y$, M=Mg, Si, Ca, Zr, La, Ce, Nd), $SiO_2$ was the most effective support for the synthesis of methanol (*14*). The methanol formation rate normalized to the total amount of Pd atoms over the Pd sulfide supported on $SiO_2$ was much higher than that over the bulk sulfide, whereas the space-time yields of methanol were comparable with each other. Considering the previous studies with the reduced Pd catalysts mentioned above, we have investigated the effect of modification of Pd sulfide supported on $SiO_2$ with several additives such as alkali metal, alkaline earth metal and/or lanthanide elements on improving its activity. Methanol synthesis activity of the supported Pd sulfide has been compared with those of the reduced Pd catalyst, the commercial $Cu/ZnO/Al_2O_3$ catalyst and/or the Mo sulfide-based catalyst in the presence and absence of $H_2S$. The formation mechanism of methanol on the supported Pd sulfides has been also examined by means of several spectroscopic techniques such as *in-situ* diffuse reflectance FTIR spectroscopy.

# Experimental

### Preparation of oxide precursors

Commercially available $SiO_2$ powder (BET surface area: 560 $m^2$ $g^{-1}$, pore volume: 0.25 mL $g^{-1}$, particle size: $150\text{-}250\times10^{-6}$ m) was used as the support in this study. Pd oxide supported on $SiO_2$ was prepared by the incipient wetness method, where precalcined $SiO_2$ powder was impregnated with an aqueous $Pd(NH_3)_4(NO_3)_2$ solution followed by drying and calcinations. Modification with additives was then carried out by impregnating thus prepared $Pd/SiO_2$ with an aqueous solution containing one of metal nitrates selected from $M(NO_3)_x$ (M=Li, K, Cs, Mg, Ca, Ba, Sr, Sc, Y, La, Ce, Nd, Mn, Zn and Al), followed by drying. Pd loading of these oxide precursors was fixed at 0.045 on the $SiO_2$ weight basis (as metal), while the atomic ratio of M to Pd was varied from 0.1 to 2.0. More detailed procedures for the preparation of the oxide precursor can be found in our previous paper (*15*).

## Sulfiding pretreatment and CO hydrogenation

Both the sulfiding pretreatment and the CO hydrogenation were carried out in the fixed bed reactor equipped with two on-line gas chromatographs. The oxide precursor was sulfided before the reaction in a stream of $H_2S$ 5%/$H_2$ 95% at 673 K and 0.1 MPa. The cumulative amount of $H_2S$ fed during the pretreatment was 160 mol-$H_2S$ mol-$Pd^{-1}$. After the sulfidation, syngas (CO 33%/$H_2$ 62%/Ar) was continuously fed to the reactor. Ar was used as an internal standard to determine accurate CO conversion and product formation rates. Typical reaction conditions were 613 K, 5.1 MPa and 20 $m^3$ (STP) kg-$cat^{-1}$ $h^{-1}$. Activity measurements were also performed with reduced oxide precursors, a homemade $ZnO/Cr_2O_3$ and/or the commercial $Cu/ZnO/Al_2O_3$ (supplied by ICI Corp.) catalysts as references. A sulfur-free apparatus was used for these catalysts. Syngas containing $CO_2$ (CO 30%/$CO_2$ 5%/$H_2$ 60%/Ar) was used for the activity measurements with the $Cu/ZnO/Al_2O_3$ catalyst. Unless otherwise stated, the reaction was carried out in the absence of $H_2S$.

## Pd K-edge EXAFS measurement

Pd K-edge EXAFS spectra of the sample after the CO hydrogenation were measured with a high-pressure chamber made of stainless steel (*16,17*). Finely powdered precursor was pressed into a self-supporting wafer, which was set in the chamber. The precursor was then sulfided in the stream of $H_2S$ 5%/$H_2$ 95% at 673 K and 0.1 MPa. Following the sulfidation, the sample was exposed to the stream of syngas (CO 33%/$H_2$ 62%/Ar) at 613 K and 5.1 MPa for 8 h. Hereafter, the syngas in the chamber was flushed with $H_2$ stream at room temperature followed by the measurement of EXAFS spectrum. EXAFS measurement was made at room temperature in a transmission mode near Pd K-edge on an EXAFS2000 laboratory system (Rigaku).

## *In-situ* diffuse reflectance FTIR measurement

Surface species formed on the sample during the CO hydrogenation was observed by the diffuse reflectance FTIR (DRIFT) spectroscopy with a high-pressure chamber equipped with ZnSe window (Spectra-Tech Inc.) (*18-23*). After the sulfidation of the precursor in the same manner as described above, the sample was exposed to the syngas (CO 33%/$H_2$ 62%/Ar) stream at 613 K and 5.1 MPa. DRIFT spectra of the sample were then measured under these conditions on an FTS6000 FTIR spectrometer (Varian) with a resolution of 4 $cm^{-1}$ and an accumulation of 128.

## Results and discussion

### Effects of the additives

In the previous studies with the reduced Pd catalysts, Gotti and Prins (*2*) investigated methanol synthesis activity of reduced Pd/$SiO_2$ modified with several additives (M=Li, Na, K, Rb Cs, Mg, Ca, Sr, Ba, Mn, Zn and La). They found that most of these additives promote methanol synthesis activity of reduced Pd/$SiO_2$, whereas rate enhancement for the methanol formation (the ratio of methanol formation rates between reduced M/Pd/$SiO_2$ and reduced Pd/$SiO_2$) was greatly different among the additives. Ca additive was the most effective for methanol formation. Considering these previous studies, we investigated at first CO hydrogenation activity and selectivity of sulfided Pd/$SiO_2$ modified with several additives.

Figure 1 shows the formation rate of products over sulfided M/Pd/$SiO_2$ (M=Li, K, Cs, Mg, Ca, Sr, Ba, Sc, Y, La, Ce, Nd, Mn, Zn, Al) at 613 K, 5.1 MPa and 20 $m^3$ (STP) kg-cat$^{-1}$ h$^{-1}$. The atomic ratio of M to Pd is 0.5 for all supported sulfides shown in this figure. The formation rate of products over the bulk sulfide is also included in this figure. One should also bear in mind that the height of each bar in this figure is equivalent to the rate of CO conversion over the supported sulfides. Most of sulfided M/Pd/$SiO_2$ show higher CO conversion rates than the sulfided Pd/$SiO_2$. Methanol is a main product with sulfided M/Pd/$SiO_2$ except the supported sulfides modified with Li, Cs and Al. Hydrocarbons (mainly methane), dimethylether and $CO_2$ are also formed on several supported sulfides. Methanol formation rate is 3-4 times higher when the supported sulfide is modified with Ca, Sr, Sc, Y, La, Ce and/or Nd. Ca additive is the most effective for methanol synthesis among them. Methanol formation rate over sulfided Ca/Pd/$SiO_2$ is 100 times higher than that over the bulk sulfide. We further investigated methanol formation rate over reduced M/Pd/$SiO_2$ (M=Li, Cs, Mg, Ca, Y, Mn, M/Pd=0.5 atom atom$^{-1}$) at 573 K, 5.1 MPa and 20 $m^3$ (STP) kg-cat$^{-1}$ h$^{-1}$, and compared them with those over supported sulfides at the same reaction conditions. Although reduced M/Pd/$SiO_2$ showed a higher methanol formation rate than the corresponding sulfide, the rate enhancements of the additives for the methanol formation were comparable with each other.

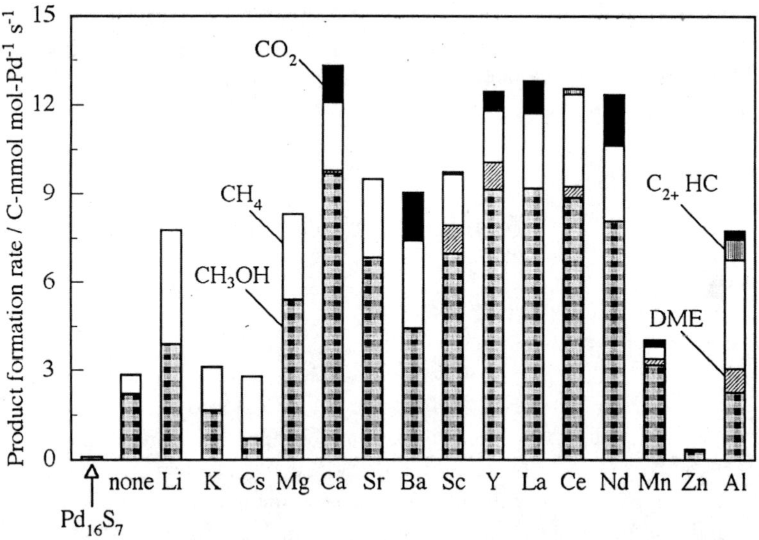


*Figure 1. Formation rate of products with the supported Pd sulfides modified with or without additives in comparison with the bulk sulfide ($P_{16}S_7$).*

The effect of the additives was then investigated in more detail. Figure 2 shows the formation rate of products with the sulfided $Ca/Pd/SiO_2$ with different atomic ratios of Ca to Pd at different reaction temperatures (613 K (a) and 573 K (b)). The reaction pressure and gas-hourly space velocity were 5.1 MPa and 20 $m^3$ (STP) kg-cat$^{-1}$ h$^{-1}$, respectively. At 613 K, methanol formation rate increases sharply with the atomic ratios of Ca to Pd up to 0.5, and shows a plateau above this ratio. The formation rates of methane and $CO_2$, on the other hand, show the maximum at Ca/Pd=0.5. At this ratio, the rate enhancement for methanol formation (4.5) was higher than that for methane formation (3.5). Thus, Ca additive could promote the formation of methanol rather than methane. It should also be noted here that sulfided $Ca/SiO_2$ showed no activity for CO hydrogenation. At the lower reaction temperature, the rate enhancements for methanol and methane formations were 5.5 and 2.5, respectively (Ca/Pd=0.5). The promoting effect of Ca additives on methanol formation was more notable at the lower reaction temperature. Similar results were obtained with supported sulfides modified with Mg and/or Y additives, whereas Cs and Mn additives hardly affected methanol formation irrespective of their contents.

The product selectivity on the sulfided $Ca/Pd/SiO_2$ (Ca/Pd=0.5) is then compared with that on the sulfided $Pd/SiO_2$ in Figure 3. The reactions were carried out at 613 K and 5.1 MPa with various gas-hourly space velocities. At the lower CO conversion (ca. 1%), methanol is a main product with the sulfided $Pd/SiO_2$ (Figure 3). As could also be observed from the figure, methane is also formed at this conversion, whereas $CO_2$ is detected at the conversion above 3%.

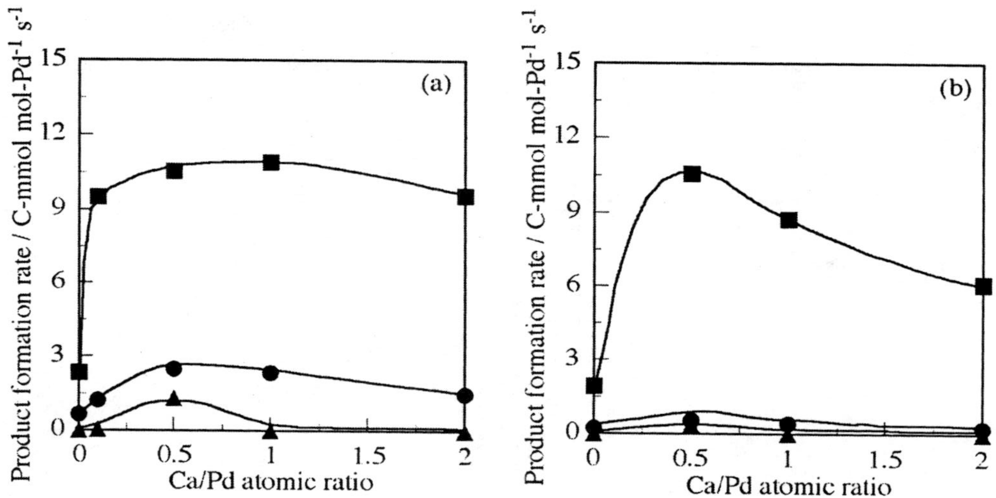


*Figure 2. Formation rate of products with the sulfided Ca/Pd/$SiO_2$ with different atomic ratios of Ca to Pd at 613 K (a) and 573 K (b). (■: methanol, ●: methane and ▲: $CO_2$).*

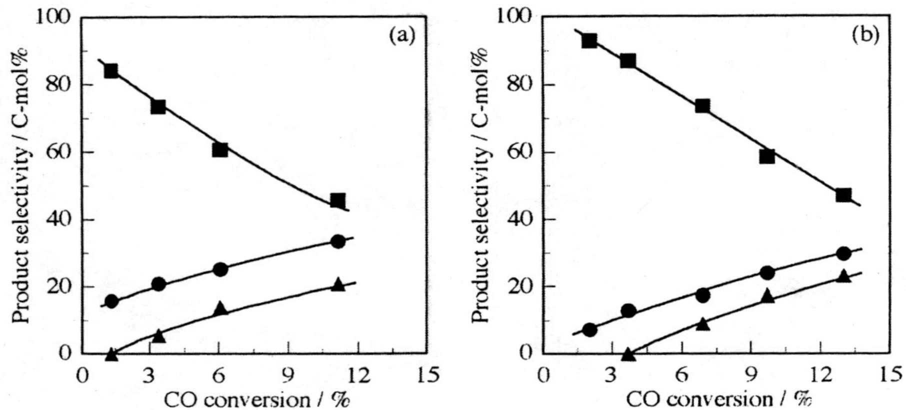


*Figure 3. Comparison of the product selectivity on the sulfided Pd/$SiO_2$ (a) and the sulfided Ca/Pd/$SiO_2$ (b). (■: methanol, ●: methane and ▲: $CO_2$).*

Thus, $CO_2$ is a secondary product. With increasing the conversion, the selectivity for methane and $CO_2$ increase at the expense of the methanol selectivity. Similar conversion-selectivity profile is obtained with sulfided Ca/Pd/$SiO_2$ (Figure 3).

The methanol selectivity on sulfided Ca/Pd/$SiO_2$ is, however, obviously higher than that on sulfided Pd/$SiO_2$. The Ca additive improves methanol selectivity as well as methanol formation rate.

## Comparison with the reduced Pd/$CeO_2$ and the Cu/ZnO/$Al_2O_3$ methanol synthesis catalysts

In this paragraph, methanol synthesis activity of sulfided Ca/Pd/$SiO_2$ is compared with those of reduced Pd catalyst, commercial methanol synthesis catalysts and/or $MoS_2$-based catalyst. Figure 4 (a) shows the STY of methanol obtained with sulfided Ca/Pd/$SiO_2$ (Ca/Pd=0.5) as a function of reaction temperature.

Dashed lines indicate equilibrium methanol STYs at each reaction temperature with gas-hourly space velocities of 30 $m^3$ (STP) kg-cat$^{-1}$ h$^{-1}$ (····) and 6 $m^3$ (STP) kg-cat$^{-1}$ h$^{-1}$ (----). At 30 $m^3$ (STP) kg-cat$^{-1}$ h$^{-1}$, the STY of methanol with sulfided Ca/Pd/$SiO_2$ shows the maximum at 593 K. From Figure 4, the STY of methanol at this reaction temperature is 730 g kg-cat$^{-1}$ h$^{-1}$. Among the reduced Pd catalysts reported so far, reduced Pd/$CeO_2$ catalyst shows the highest methanol synthesis activity, especially at the lower reaction temperatures (*6*). According to the literature, this catalyst yields 310 g kg-cat$^{-1}$ h$^{-1}$ of methanol at 473 K and 2.0 MPa. On the other hand, the STY of methanol with sulfided Ca/Pd/$SiO_2$ is only 60 g kg-cat$^{-1}$ h$^{-1}$ at the same reaction temperature (Figure 4 (a)). The STY of methanol with supported sulfide is no more than 20% of that with reduced catalyst. However, since Pd content of reduced Pd/$CeO_2$ catalyst is 3.7 times higher than that of sulfided Ca/Pd/$SiO_2$, the methanol STY over supported sulfide normalized to the total Pd atoms reaches 75% of that with reduced catalyst. Figure 4 (a) also shows the results obtained with the homemade ZnO/$Cr_2O_3$ and the commercial Cu/ZnO/$Al_2O_3$ catalysts. The ZnO/$Cr_2O_3$ catalyst was also used here because this type of catalyst shows a superior sulfur resistance than the Cu/ZnO-type catalyst (*12*). At 593 K and 30 $m^3$ (STP) kg-cat$^{-1}$ h$^{-1}$, the STY of methanol with the ZnO/$Cr_2O_3$ catalyst is much lower than that with the supported sulfide. For the commercial Cu/Zn/Al catalyst, the feed containing both the syngas and $CO_2$ was used here because it is well known that small amounts of $CO_2$ in the feed greatly improves its methanol synthesis activity (*24*). At 593 K and 30 $m^3$ (STP) kg-cat$^{-1}$ h$^{-1}$, the STY of methanol with Cu/ZnO/$Al_2O_3$ catalyst is much higher than that with supported sulfide, and reaches the equilibrium value. At 523 K and 6 $m^3$ (STP) kg-cat$^{-1}$ h$^{-1}$ where the Cu/ZnO/$Al_2O_3$ catalyst is usually used, methanol STY is 1200 g kg-cat$^{-1}$ h$^{-1}$, which is still higher than that with supported sulfide.

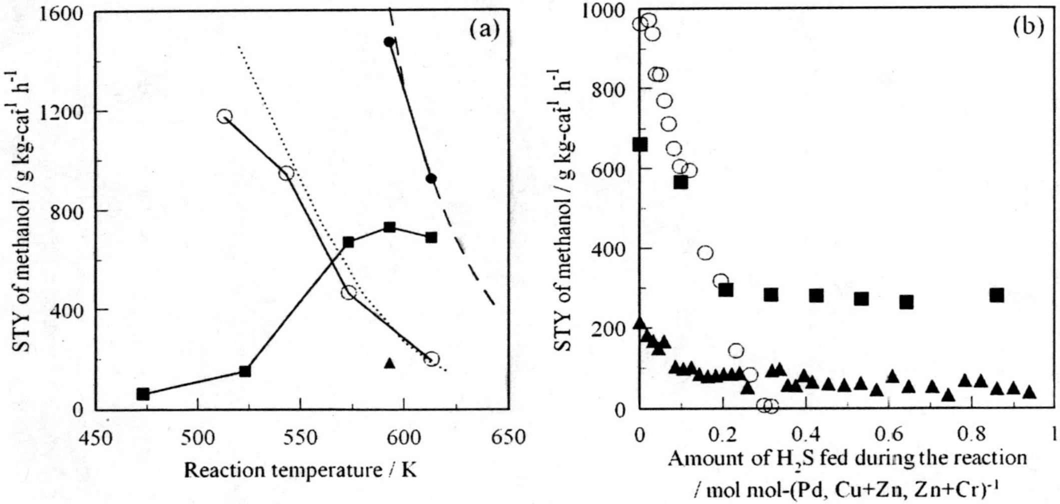


*Figure 4. Space-time yield of methanol with the sulfided Ca/Pd/$SiO_2$ (■) in the absence (a) and presence (b) of $H_2S$ in comparison with the ZnO/$Cr_2O_3$ (▲) and the commercial Cu/ZnO/$Al_2O_3$ catalysts (○, ●). (open symbols: 6 $m^3$ (STP) kg- cat$^{-1}$ $h^{-1}$, closed symbols: 30 $m^3$ (STP) kg-cat$^{-1}$ $h^{-1}$)*

Then, effects of $H_2S$ were examined. After methanol synthesis activities of each catalyst reached steady states under the sulfur-free conditions, syngas mixed with 0.1% $H_2S/H_2$ was fed to the reactor. Methanol STYs as a function of the total amount of $H_2S$ fed during the reactions are shown in Figure 4 (b). The amount of $H_2S$ fed during the reaction in this figure is normalized to the metal content (Pd, Cu+Zn or Zn+Cr) of each catalyst. Soon after the syngas containing $H_2S$ ($H_2S$ concentration: 120 ppm) was fed into the reactor, the STY over sulfided Ca/Pd/$SiO_2$ decreases to 35% of the initial value. Thereafter, this catalyst exhibits a constant STY, *i.e.* 250 g kg-cat$^{-1}$ h$^{-1}$ even when the total amount of $H_2S$ reaches 6.0 mol-$H_2S$ mol-Pd$^{-1}$. In marked contrast to the supported sulfide, the STYs of methanol with ZnO/$Cr_2O_3$ and commercial Cu/ZnO/$Al_2O_3$ catalysts decrease with increasing total amounts of $H_2S$ fed during the reaction. Especially, the STY over the Cu/ZnO/$Al_2O_3$ catalyst decreases much more rapidly compared with ZnO/$Cr_2O_3$ catalyst and eventually drops into 0 at about 0.3 mol-$H_2S$ mol-(Cu+Zn)$^{-1}$. In other words, sulfided Ca/Pd/$SiO_2$ have much superior sulfur tolerances than ZnO/$Cr_2O_3$ and Cu/ZnO/$Al_2O_3$ catalysts. It is also worthy to note that Mo sulfide modified with K additive is reported to yields 230 g kg-cat$^{-1}$ h$^{-1}$ of methanol at 553 K and 21 MPa in the presence of $H_2S$ (ca. 50 ppm) (*11*). A comparable STY of methanol can be obtained with sulfided Ca/Pd/$SiO_2$ in the presence of $H_2S$ even at 5.1 MPa. In the presence of $H_2S$, it can be said that sulfided Ca/Pd/$SiO_2$ shows the highest methanol synthesis activity among the catalysts reported so far.

## Role of the additive

As mentioned in the above section, the present study found that the rate enhancement for methanol formation over the sulfided M/Pd/$SiO_2$ is comparable with that over the corresponding reduced M/Pd/$SiO_2$ (M= M=Li, Cs, Mg, Ca, Y, Mn, M/Pd=0.5 atom atom$^{-1}$). This implies that the role of these additives in methanol formation over supported sulfides is similar to that over reduced catalysts. In the previous studies with methanol synthesis over reduced Pd catalyst, however, there exist different speculations about the role of the additives and/or the support. On one hand, Ryndin *et al.* (*4*) claimed that the lanthanide oxide affects the morphology of the metallic Pd species when used as the support. On the other hand, several researchers claimed that the additives such as Mg and Ca stabilize the reaction intermediate for the methanol formation (*1-3,8,9*). Furthermore, we find a discrepancy among the previous studies as concerns the reaction intermediate, *i.e.* formate species vs. formyl species. Such the discrepancies may arise from lack of information about the structure of Pd species and the surface species formed during methanol synthesis, especially under the high-pressure reaction conditions. To examine the role of the additive in methanol synthesis over supported sulfides, they were investigated under high-pressure methanol synthesis conditions by EXAFS and DRIFT spectroscopies.

*Structure of Pd species after the methanol synthesis*

First, the structure of Pd species on sulfided M/Pd/$SiO_2$ after methanol synthesis was investigated by Pd K-edge EXAFS, and compared with that on sulfided Pd/$SiO_2$. Pd K-edge EXAFS spectra were measured after methanol synthesis at 613 K and 5.1 MPa for 8 h without contact with air. Fourier transforms of EXAFS spectra of both sulfided Ca/Pd/$SiO_2$ (Ca/Pd=0.5) and sulfided Pd/$SiO_2$ showed Pd-S and Pd-Pd peaks and were similar to each other (not shown here). From the comparison with some reference compounds such as PdS, $Pd_{16}S_{17}$, $Pd_4S$ and Pd metal, it was suggested that Pd species exists as $Pd_{16}S_7$ or $Pd_4S$ on both supported sulfides. XRD measurements (in air) of sulfided M/Pd/$SiO_2$ (M=Cs, Mn, Mg and Ca, M/Pd=0.5) and sulfided Pd/$SiO_2$ after methanol synthesis further indicated the formation of $Pd_4S$ rather than $Pd_{16}S_7$. This result is consistent with the thermodynamic calculation for the gas-solid equilibrium of the Pd-$H_2S$/$H_2$ system where $Pd_4S$ is a stable phase in the presence of $H_2S$ above 0.35 ppm in concentration at 613 K, because the apparatuses for the activity and EXAFS measurements were contaminated with $H_2S$ ca several ppm in concentration even though the sulfur-free conditions. XRD patterns of these supported sulfides also indicated that the crystalline size of $Pd_4S$ is not affected by the modification with Cs, Mn, Mg and/or Ca additives.

*Observation of surface species formed during the methanol synthesis*

Then, the surface species were investigated under high-pressure methanol synthesis conditions by DRIFT spectroscopy. DRIFT spectra were recorded as a function of time when the sulfided Ca/Pd/$SiO_2$ (Ca/Pd=0.5) was exposed to the stream of syngas at 613 K and 5.1 MPa (Figure 5 (a)).

In addition to the bands assigned to the gas-phase methanol (1059, 1032, 1001 $cm^{-1}$), several bands are clearly observed in the spectra. To facilitate the assignment of these bands, the reaction temperature was reached down to 298 K and then the chamber was flushed with helium flow. Although the bands for the gas-phase products completely disappear, the bands at 1600 and 1270 $cm^{-1}$ are still visible. Several authors have reported that IR bands of adsorbed formate species are observed when reduced Pd catalysts are exposed to the stream of syngas at elevated temperatures (*25,26*). The formate species yields IR bands at 1600-1550 $cm^{-1}$, 1400-1380 $cm^{-1}$, 1380-1340 $cm^{-1}$ and 1090-1060 $cm^{-1}$ (*27*). The band at 1600 $cm^{-1}$ shown in Figure 5 (a), therefore, can be assigned to the formate species. It should be noted that the spectra show no IR bands at around 1700 $cm^{-1}$ although the band assigned to formyl species appears at 1748 $cm^{-1}$ when reduced Ca/Pd/$SiO_2$ is exposed to methanol vapor at room temperature (*28*). The appearance of the band at 1450 $cm^{-1}$ also suggests the formation of methoxy or methyl species. The same protocol was applied to other supported

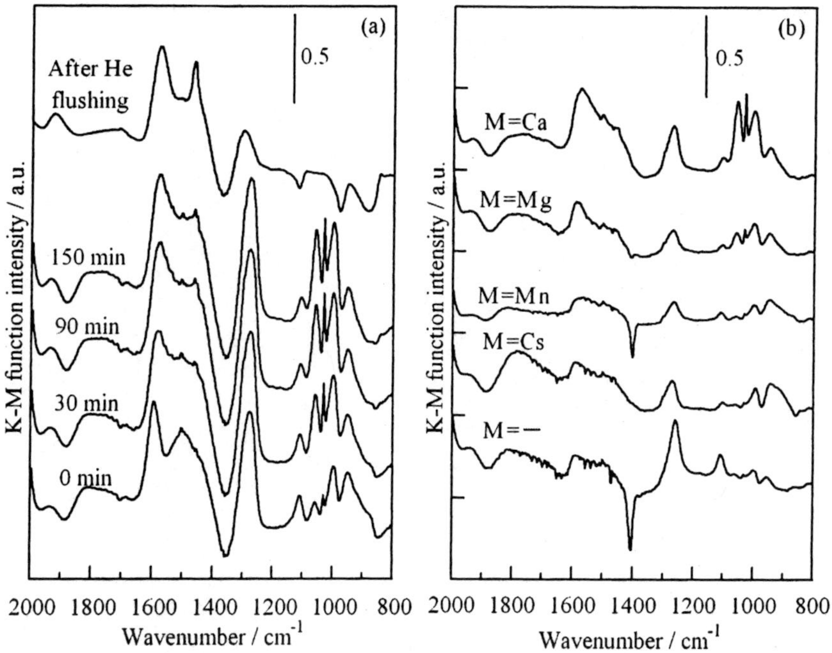


*Figure 5. DRIFT spectra of the sulfided Ca/Pd/$SiO_2$ during the methanol synthesis at 613 K and 5.1 MPa as a function of time on stream (a). Effect of different additives on the DRIFT spectra of sulfided Pd/$SiO_2$ during the methanol synthesis at 150 min on stream is shown in (b).*

sulfides. Figure 5 (b) shows DRIFT spectra of sulfided M/Pd/$SiO_2$ (M=Cs, Mn, Mg, Ca, M/Pd=0.5) and sulfided Pd/$SiO_2$ during methanol synthesis. These spectra were measured when supported sulfides were exposed to syngas stream at 613 K and 5.1 MPa for 150 min. IR bands for the gas-phase methanol are clearly observed in the spectra of the supported sulfides modified with Mg and/or Ca additives, whereas only trace bands for the gas-phase methanol appear in the spectra for other supported sulfides. This trend is consistent with that observed for their methanol synthesis activities shown in Figure 1. Furthermore, the intensity of the band for the formate species (ca. 1600 $cm^{-1}$) changes depending on the additives. The intensity of the band for the gas-phase methanol is stronger when the intensity of the band for the formate species is stronger. In Figure 6 (a), the intensity of the band for the gas-phase methanol (1032 $cm^{-1}$) is plotted against that for the formate species. As can be seen from this figure, the intensity of the band for the gas-phase methanol increases with increasing the intensity of the formate band. This simple relationship suggests that the formate

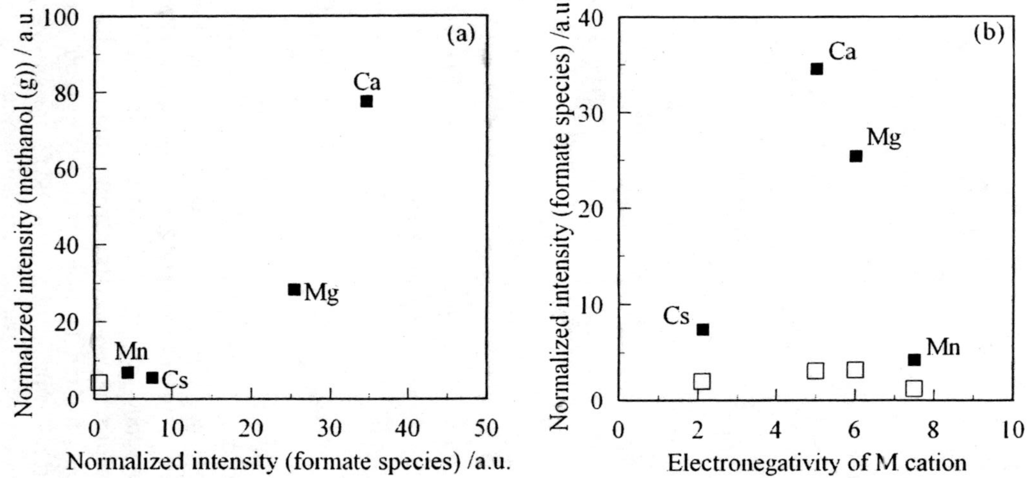


*Figure 6. Relationship between the intensity of the band for the formate species and the intensity of the band for the gas-phase methanol (a), or the electro-negativity of M cations (b).(Open symbols: sulfided M/$SiO_2$, closed symbols: sulfided M/Pd/$SiO_2$ with the atomic ratio of M to Pd of 0.5).*

species is a reactive intermediate for methanol synthesis over supported sulfides and the hydrogenation of the formate species is a rate determing step in methanol formation.

Concerning the stabilization of the formate species, Gotti and Prins *(2)* found a volcano-type relationship between the rate enhancement for methanol formation over reduced $M/Pd/SiO_2$ catalysts and the electro negativity of the additive cations with the peak at the Ca cation. Based on this result, they proposed that Ca cation stabilizes the formate species moderately, which reacts with hydrogen atoms provided with the metallic Pd sites to form methanol effectively. According to their proposal, the formate species is stabilized too strongly on the cations with lower electronegativity than the Ca cation to react with hydrogen atoms. Contrary to their proposal, a volcano-type relationship with the peak at the Ca cation is observed when the intensity of the band for the formate species in Figure 6 (a) is plotted as a function of the electronegativity of the additive cations (Figure 6 (b)). We further investigated the intensity of the formate species observed in the DRIFT spectra of the sulfided $M/SiO_2$ (M= Cs, Mn, Mg, Ca) during exposing the samples to syngas stream at 613 K and 5.1 MPa. Their intensities are also plotted as a function of the electronegativity of the M cation in Figure 6 (b). As can be seen from this figure, the intensity of the formate species is much weaker for the sulfided $M/SiO_2$. These results clearly indicate that the Pd sulfide is indispensable for the stabilization of the formate species as well. In other words, both the Pd sulfide species and the Ca additive are involved in the stabilization of the formate species, which results in effective methanol formation over the supported sulfide.

## Reference


1. Naito, S.; Yoshioka, H.; Orita, H.; Tamaru, K. *Proc. 8th Int. Congr. Catal.*, **1984**, *2*, 163.
2. Gotti, A.; Prins, R. *J. Catal.*, **1998**, *175*, 302.
3. Driessen, J. M.; Poels, E. K.; Hindermann J. P.; Ponec, V. *J. Catal.*, **1983**, *82*, 26.
4. Ryndin, Y. A.; Hicks, R. F.; Bell, A. T. *J. Catal.*, **1981**, *70*, 287.
5. J. S. Rieck and A. T. Bell, *J. Catal.*, **99**, 278 (1986).
6. Matsumura, Y.; Shen, W.-J.; Ichihashi, Y.; Okumura, M. *J. Catal.*, **2001**, *197*, 267.
7. Kelly, K.P.; Tatsumi, T.; Uematsu, T.; Driscoll, D. J.; Lunsford, J. H. *J. Catal.,* **1986**, *101*, 396.
8. Lee, G. V. D.; Ponec, V. *Catal. Rev.-Sci. Eng.*, **1987**, *29 (2&3)*, 183.
9. Hindermann, J. P.; Kiennemann, A.; Chakor-Alami, A.; Kieffer, R. Proc. *8th Int. Congr. Catal.*, **1983**, *2*, 163.
10. Koizumi, N.; Miyazawa, A.; Furukawa, T.; Yamada M. *Chem. Lett.,* **2001**, 1282.

11. Quarderer, G. J.; Cochran, G. A. *European Patent*, **1984**, *119*, 609.
12. Wood, B. J.; Isakson, W. E.; Wise, H.; *Ind. Eng. Chem. Prod. Res. Dev.*, **1980**, *19*, 197.
13. Berube, M. N.; Sung, B.; Vannice, M. A.; *Appl. Catal.*, **1987**, *31*, 133.
14. Koizumi, N.; Murai, K.; Ozaki, T.; Yamada, M. *Catal. Today*, **2004**, *89*, 465.
15. Koizumi, N.; Murai, K.; Tamayama, S.; Ozaki, T.; Yamada, M. *Energy & Fuels*, **2003**, *17 (4)*, 829.
16. Koizumi, N.; Iijima, M.; Mochizuki, T.; Yamada M. *Stud. Surf. Sci. Catal.*, **1997**, *106*, 293.
17. Koizumi, N.; Yamazaki, M.; Hatanaka, S.; Yamada M. *Catal. Today*, **1997**, *39*, 33.
18. Jiang, M.; Koizumi, N.; Yamada M. *J. Phys. Chem. B*, **2000**, *104*, 7636.
19. Jiang, M.; Koizumi, N.; Yamada M. *Appl. Catal. A: General*, **2000**, *204*, 49.
20. Jiang, M.; Koizumi, N.; Ozaki, T.; Yamada M. *Appl. Catal. A: General*, **2000**, *209*, 59.
21. Bian, G.; Oonuki, A.; Kobayashi, Y.; Koizumi, N.; Yamada M. *Appl. Catal. A; General*, **2001**. *219*, 13.
22. Bian, G.; Nanba, T.; Koizumi, N.; Yamada M. *J. Mol. Catal. A: Chemical*, **2002**, *178*, 219.
23. Bian, G.; Oonuki, A.; Koizumi, N.; Nomoto H.; Yamada M. *J. Mol. Catal. A: Chemical*, **2002**, *186*, 203.
24. Zhang, Y.; Sun, Q.; Deng, J.; Wu, D.; Chen, S. *Appl. Catal. A: General*, **1997**, *158*, 105.
25. Kikuzono, Y.; Kagami, S.; Naito, S.; Onishi, T.; Tamaru, K. *Faraday Discuss. Chem. Soc.*, **1981**, *72*, 135.
26. Anderson, J. A.; Lopez-Granados, M.; Fernandez-Garcia, M. *J. Catal.*, **1998**, *176*, 235.
27. Gopal, P. G.; Schneider, R. L.; Watter, K. L. *J. Catal.*, **1987**, *105*, 366.
28. Cabilla, G. C.; Bonivardi, A. L., Baltanas, M. A. *J. Catal.*, **2001**, *201*, 213.

## Chapter 11

# The Effect of Continuous $H_2S$ Exposure on the Performance of Thick Palladium–Copper Alloy Membranes

**Bret H. Howard[1], Anthony V. Cugini[1], Richard P. Killmeyer[1], Bryan D. Morreale[2], and Robert M. Enick[3]**

**[1]United States Department of Energy, National Energy Technology Laboratory (NETL), 626 Cochrans Mill Road, Pittsburgh, PA 15236**
**[2]NETL Support Contractor, Parsons, P.O. Box 618, South Park, PA 15129**
**[3]NETL Research Associate, Department of Chemical and Petroleum Engineering, University of Pittsburgh, Pittsburgh, PA 15260**

Membranes fabricated from Pd-Cu alloys containing 80, 60 and 53wt%Pd, as well as pure Pd, were exposed to flowing 1000 ppm $H_2S$ in $H_2$ over the temperature range of 350 to 900°C using three approaches to verify NETL's previously reported transient $H_2S$ exposure results. 100 μm thick braze-mounted foils failed prior to 600°C due to apparent sulfur attack at the braze. 1000 μm thick welded membranes demonstrated similar trends as found using the transient method in that hydrogen flux through the Pd-Cu alloys with fcc structure was not significantly degraded by $H_2S$ exposure. However, both of these experimental methods suffered from possible disadvantages. The transient method had limited $H_2S$ availability and limited exposure duration, and in the 1000 μm steady-state tests, bulk diffusion limitations could mask effects resulting from $H_2S$ exposure. Preliminary results obtained using an alternative membrane mounting method and test protocol for steady-state testing of 100 μm thick Pd and 80wt%Pd-Cu foils at 350°C showed that significant flux losses occurred on exposure to flowing 1000 ppm $H_2S$ in $H_2$, contrary to the earlier studies. Characterization showed that relatively thick sulfide layers had developed on the membrane surfaces during the 120 hours of exposure.

## Introduction

The U.S. Department of Energy's National Energy Technology Laboratory (NETL) is currently engaged in the investigation of membrane materials for hydrogen separation with a specific interest in contaminant resistant metal alloy membrane materials. This interest is based on the view that fossil fuel resources, particularly coal, will play a major role in the transition to a hydrogen economy by providing a near- to mid-term source of hydrogen[1]. Advanced energy plants, such as those promoted by the U.S. DOE's *FutureGen* initiative, can utilize domestic fossil fuels much more cleanly than current technologies[2], and, when coupled with carbon dioxide sequestration to mitigate the environmental concerns associated with global climate change, such plants could produce essentially emission-free energy.

Coal, through technologies based on gasification combined with water-gas shift (WGS), can be used to produce hydrogen mixed with carbon dioxide and trace amounts of other gases. Membranes are one of the promising technologies for the separation of hydrogen from other product gases; however, the materials comprising the membrane must be able to endure the harsh gasification environment. Palladium-based membranes and membranes utilizing a palladium-based catalytic layer on a hydrogen permeable substrate are potential candidates because of their high hydrogen permeability and catalytic activity with respect to hydrogen dissociation. However, a significant technical barrier impeding hydrogen separation membrane development is the susceptibility of palladium-containing membranes to poisoning or corrosion by components present in gasifier effluent streams. Probably of most concern is the effect of hydrogen sulfide, which can be present at concentrations as high as ~2% in a raw syngas stream down to low ppm levels following desulfurization processes.

The impact of hydrogen sulfide and other sulfur-containing compounds on membranes based on palladium or having a palladium component is well known and has often been reported in the literature [3-13]. Such reports cover a wide range of membrane compositions, gas compositions and exposure conditions and the reported effect of sulfur on membrane performance varies widely. Therefore, the identification of sulfur tolerant membrane materials and the understanding of the mechanisms of poisoning are critical to the development of technologies dependent on hydrogen membranes. Recently, promising developments have been realized in enhancing the sulfur-resistance of metal membranes. For example, resistance to sulfur poisoning upon exposure to methyl disulfide has been reported for thin Pd films formed within the macropores of an alumina support[14] and the application of sub-micron films of Pt on Pd has been suggested to enhance sulfur resistance of hydrogen membranes[5, 15, 16].

Some of the most promising investigations of sulfur resistant membranes have focused on Pd alloys, especially Pd-Cu alloys. The interest in Pd-Cu alloy membranes is a result of two promising material characteristics. First, specific Pd-Cu alloys, 60wt%Pd-40wt%Cu, have shown flux values slightly greater than pure palladium at 350°C[4]. Second, Pd-Cu alloys have been reported to exhibit "*resistance*" to sulfur poisoning in hydrogen-rich streams containing $H_2S$ concentrations up to 5 ppm[14], 1000 ppm[4] and intermittent exposure to 100,000 ppm[5]. Although McKinley reported "resistance" to permanent poisoning in the presence of $H_2S$ for the Pd60wt%-40wt%Cu alloy at 350°C, significant flux decreases were observed during exposure to concentrations as low as 4 ppm $H_2S$ at 350°C[4].

NETL has been actively investigating through both experimental and computational studies the potential of Pd-Cu alloys because of their potential applicability for gasifier and post-gasifier water-gas shift (WGS) membrane reactors and similar harsh environment hydrogen separation applications. Recent computational studies by Sholl and Alfonso have focused on the prediction of alloy membrane permeability values and the interaction of S with potential membrane materials[17-19]. Our experimental studies on the permeability of a series Pd-Cu alloys in pure hydrogen and in the presence of $H_2S$ have also been recently reported[20,21].

The performance of 100-μm thick 100, 80, 60, and 53wt%Pd membranes in $H_2$ was reported by our group[20]. Relatively thick membranes, rather than ultra-thin micron-scale films, were selected in order to facilitate fabrication, ensure precise composition, and retain mechanical integrity. In summary, at temperatures below ~500°C corresponding to the bcc/mixed bcc-fcc stability range, the 60wt%Pd alloy exhibited the highest overall permeability of any of the tested alloys, about 70% of the permeability of Pd at 350°C. This permeability was slightly lower than some reported permeability values. Possible causes for the lower permeability include intermetallic diffusion at the membrane edge resulting from welding and alloy impurities such as carbon. The performance dropped approximately an order of magnitude above this temperature corresponding to the conversion to fcc. The 80wt%Pd alloy, fcc over the entire temperature range studied, exhibited permeability values of about 30% of Pd, however, it exhibited higher permeability values than the 53wt% Pd alloy (bcc or fcc structured) at all temperatures. A general trend found in this series of experiments was that when comparing fcc to fcc or bcc to bcc, a higher Pd content resulted in a higher hydrogen permeability.

Morreale et al. reported the permeability results for the same series of Pd-Cu alloys in the presence of 1000 ppm $H_2S$ in $H_2$[21]. These experiments were conducted using a transient (batch) flux measurement approach which employed a finite amount of $H_2S$ for relatively short test durations. Further, the calculation of permeability was based on the assumption that the interaction between $H_2S$ and the membrane surface was essentially instantaneous. This study showed that the 80wt%Pd alloy membrane, fcc throughout the entire temperature range of study, exhibited essentially no change in hydrogen permeability over the entire 350 to 900°C temperature range as compared to the baseline tests. For the 60wt%Pd alloy in the presence of 1000 ppm $H_2S$, the permeability essentially matched the baseline permeability above about 600°C where the alloy is fcc in structure. However, at temperatures corresponding to the mixed fcc-bcc to bcc region, below about 440°C, the permeability dropped significantly, by as much as about two orders of magnitude at 350°C. This same performance trend was found for the 53wt%Pd alloy. The results of these transient performance experiments indicated that when the Pd-Cu alloy had an fcc structure, $H_2S$ had little impact on permeability but when the structure was bcc, $H_2S$ had a moderate to severe impact.

These transient experiments strongly suggested a correlation between the alloy crystal structure and $H_2S$ tolerance. However, due to limitations of the experimental method (finite $H_2S$ availability, competition for the available $H_2S$ by other metal surfaces, relatively short test durations and the limited data set), more representative, steady-state performance experiments were deemed essential for an accurate assessment of the impact of $H_2S$ on membrane performance.

Therefore, the objective of this study was to verify the results of the transient $H_2S$ exposure study previously reported by Morreale et al.[21] using a steady-state approach similar to that used in previous NETL membrane studies[20, 22, 23]. This study utilized the same series of three Pd–Cu alloys which were selected based on the Pd-Cu phase diagram[24]. The alloys were 53wt%Pd-47wt%Cu (the composition which intersects the apex of the B2 region, commonly referred to as bcc), 60wt%Pd-40wt%Cu (the composition reported as yielding the greatest $H_2$ permeability[25]) and 80wt%Pd-20wt%Cu (a composition having an A1 structure, commonly referred to as fcc, throughout the temperature range of interest). Pd was included as a reference. The terms fcc and bcc will be used in this paper for consistency with previous work. For the remainder of this paper, the Pd-Cu alloys will be referred to by their palladium content in weight percent with the balance understood to be copper.

## Experimental Section

Palladium foils, 99.9%, Alfa Aesar, and palladium-copper alloy foils, 99.9%, ACI Alloys, were obtained in approximately 100 μm and 1000 μm thicknesses for this study. The compositions of the Pd-Cu alloy foils were 80, 60, and 53wt% Pd. The crystalline phase and composition of the alloy samples was determined by X-ray diffraction (XRD) and inductively coupled plasma atomic emission spectroscopy, respectively. All of the foils used in this study were cleaned to remove surface contamination and oxides prior to mounting. Membranes were typically cleaned by sanding with fine silicon carbide sandpaper, polishing with silicon carbide, rinsing with distilled water and then acetone. Membranes were mounted for permeability testing using three methods. The 100 μm thick foils used in the steady-state experiments were gold brazed to nickel-copper alloy supports before being welded into the testing fixture. The 1000 μm thick foils used for the continuous $H_2S$ exposures were TIG welded directly into the fixtures. The 100 μm thick foils used for the continuous $H_2S$ exposures were mounted in Swagelok® VCR fittings for testing, avoiding the need to directly weld or braze the membrane foil. All methods employed 1 mm thick porous Hastelloy® or porous 316 stainless steel support disks. The porous supports were isolated from the membrane foil by a ~100 μm thick porous alumina intermetallic diffusion barrier. Details of the weld- and braze-based procedures have been reported previously[22].

Membrane performance testing was conducted at the NETL Hydrogen Technology Research Facility utilizing the Hydrogen Membrane Testing (HMT) unit that is dedicated to in-depth studies under steady-state, continuous flow conditions using $H_2S$-containing gas streams, described in detail elsewhere[20, 22, 23]. A 90% $H_2$ -10% He (referred to as $H_2$ in this manuscript) or a 1000 ppm $H_2S$ - 10% He - balance $H_2$ (referred to as 1000 ppm $H_2S$ in this manuscript) mixture was fed to the unit on the retentate side and ultra-high purity argon sweep gas was fed to the permeate side. The argon was supplied at a rate that ensured that the $H_2$ concentration in the permeate was at about 4%. All steady-state tests reported here were carried out at a retentate total pressure of 620 kPa. Due to the high temperatures associated with mounting the membrane (brazed and welded), the crystalline structure of the alloy could not be verified prior to testing. Therefore, initial testing was conducted under $H_2$ at a temperature corresponding to the desired crystalline phase and the membrane was held at that temperature until no change in flux was observed. Once the flux stabilized at the initial temperature, equilibrium crystal structure was assumed to have been attained. Each individual membrane sample was then exposed to the test gas stream in increasing temperature steps (350 to 900°C). Each temperature was held until a steady state flux was observed before continuing to the next condition.

Membranes were characterized after the completion of a test series or following a membrane failure. The full characterization protocol consisted of the following steps. First, following removal of the membrane from the testing assembly, the retentate side surface was examined using a stereomicroscope for macro-scale changes such as color change or cracking. Next, the surface was examined by SEM/EDS to determine micro-scale morphological changes such as alloy grain growth or the crystallization of new phases. EDS was used to determine the elemental constituents of observed features. XRD was used to determine the alloy phase composition and the identity of any bulk metal sulfides or other phases present. In some cases, the membrane was cross sectioned and examined by SEM/EDS to study the internal membrane structure. Because of the approach used for this study (the membrane remained in the reactor for the complete test series), the membranes were generally not characterized until after exposure to the $H_2S$-containing test gas at 900°C.

## Results and Discussion

### Steady-state $H_2S$ exposure of 100 μm foils

Initial experiments carried out to examine the performance of the Pd-Cu alloys under continuous exposure to $H_2S$ utilized the same membrane mounting method used for previous the study conducted by Morreale et al. - - a membrane foil brazed to a support ring. The experimental protocol used was to heat the membrane to the initial test temperature of 350°C under inert gas flow. The $H_2$ test gas was then introduced in order to verify that the membrane's performance was approximately as expected based on previous performance tests. Upon verification of acceptable performance, the test gas mixture containing 1000 ppm $H_2S$ was introduced into the system. The $H_2$ flux was monitored until a steady state flux was observed. Following sufficient steady state data collection, the temperature was increased to the next test condition.

Several attempts were made to carry out this test using Pd as well as the Pd-Cu alloys. In each attempt, membrane failure occurred at a temperature between 500 and 600°C, indicated by He in the permeate stream. Following failure, samples were removed from the test assembly and examined using an optical microscope. Cracks were noted in several cases at the membrane-braze interface suggesting that failure occurred at this location. Examination of membrane cross sections by SEM/EDS revealed the presence of sulfides at the surface of and within the Au braze and also in the membrane alloy near the braze. It is suspected that the formation of relatively brittle sulfides within the membrane and braze, coupled with the mechanical stresses from sulfide growth as well as thermal stresses during testing, led to the observed membrane failures. A cross-section of a failed membrane is shown in Figure 1 illustrating the concentration of S in and near the braze.

## Steady-state $H_2S$ exposure of 1000 µm foils

Due to the 100 µm membrane structural failures linked to sulfide growth in the braze region of the membrane assembly, continuous $H_2S$ exposure testing was attempted using a new mounting approach. This approach utilized 1000 µm thick foils welded directly into the holders. The extreme thickness of the membrane foil was expected to enable the membrane to resist structural failure in the weld region for a sufficient duration for data collection. However, due to the membrane thickness, effects resulting from sulfur interaction with the membrane surface could be masked by transport limitations through the membrane bulk.

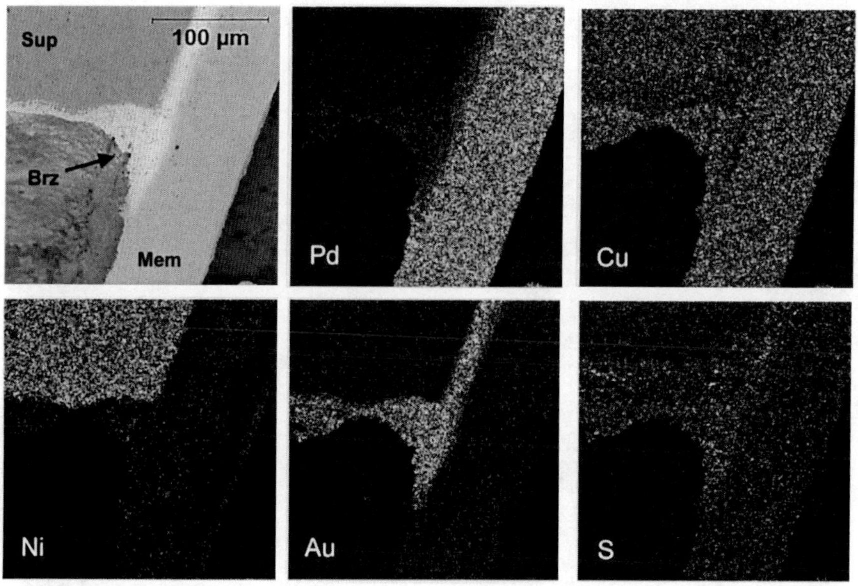


***Figure 1.*** *SEM image and X-ray maps of failed 60wt%Pd membrane cross-section. Sulfur has accumulated preferentially in the Au braze at the membrane-braze interface and appears to have migrated a substantial distance into the Ni-Cu support ring. (Sup = support ring, Mem = membrane foil, Brz = braze area)*

The experimental protocol used was generally the same as described above for the 100 µm membranes. Membranes were heated to the initial test temperature, usually 350°C, under an inert gas flow and then the $H_2$ test gas was introduced to verify that the membrane's performance was approximately as expected. Upon verification of acceptable performance, the test gas mixture containing 1000 ppm $H_2S$ was introduced into the system. The $H_2$ flux was monitored until a steady-state level was achieved. Following sufficient steady-state data collection, the temperature was increased to the next test condition. Barring membrane structural failure or system upsets, a membrane remained in the test unit until test completion, usually 900°C.

Permeability results for the 1000 μm thick 100, 80, 60 and 53wt% Pd membranes as a function of inverse absolute temperature in the presence of the flowing 1000 ppm $H_2S$ test gas are presented in Figures 2-5. The neat $H_2$ permeability values for the same 1000 μm thick alloys are also included in the figures for comparison. The $H_2$ permeability of 1000 μm thick Pd is shown in each figure for reference.

The permeability of the 100wt%Pd membrane with both $H_2$ and $H_2$/1000 ppm $H_2S$ atmospheres is shown in Figure 2. The 100wt%Pd membrane showed about a 15 to 50% drop in hydrogen permeability in the presence of 1000 ppm $H_2S$ over the 350 to 900°C temperature range, and the drop was more significant at higher temperatures. Pd has an fcc structure over this entire temperature range.

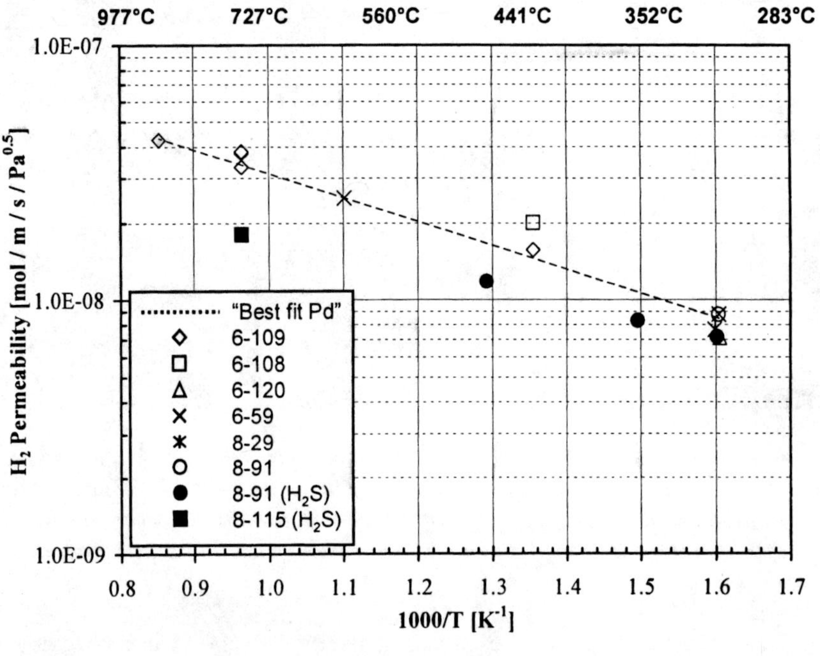


*Figure 2. Steady-state hydrogen permeability results for the 1000-μm thick 100wt%Pd membrane as a function of inverse temperature under $H_2$ and 1000 ppm $H_2S/H_2$. (Open symbols signify $H_2$ and filled symbols signify 1000 ppm $H_2S/H_2$. Hyphenated numbers in figure key are the membrane identification numbers.)*

The permeability of the 80wt%Pd alloy membrane is shown in Figure 3. The 80wt%Pd alloy membrane showed up to about a 26% drop in hydrogen permability in the presence of 1000 ppm $H_2S$ over the 350 to 900°C temperature range. The 80wt%Pd alloy structure is also fcc over this entire temperature range.

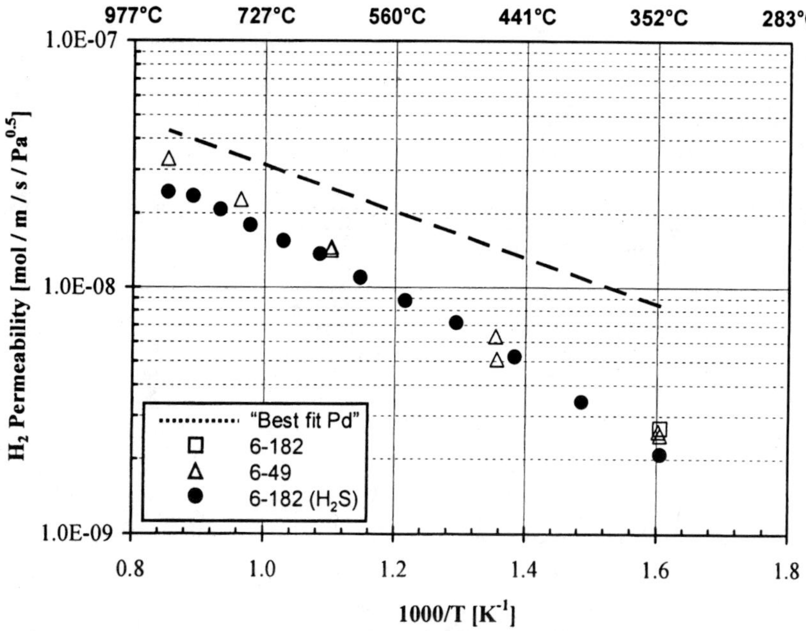


***Figure 3.*** *Steady-state hydrogen permeability results for the 1000 μm thick 80wt%Pd membrane as a function of inverse temperature under $H_2$ and 1000 ppm $H_2S/H_2$. (Open symbols signify $H_2$ and filled symbols signify 1000 ppm $H_2S/H_2$. Hyphenated numbers in figure key are the membrane identification numbers.)*

Figure 4 shows the permeability of the 60wt%Pd alloy membrane in the presence of 1000 ppm $H_2S$ over the 350 to 900°C temperature range. The permeability values showed two distinct trends when compared to the performance in clean $H_2$. At 350°C, the permeability in the presence of $H_2S$ was nearly two orders of magnitude lower than in $H_2$. As temperature increased, the permeability in the presence of $H_2S$ converged with that for $H_2$ until they were essentially equal at about 450°C. This range corresponds to the temperature stability range for mixed bcc-fcc structure. From this temperature, going through the transition to completely fcc in structure at 600°C and continuing up to 900°C, the permeability in the presence of $H_2S$ essentially matched the permeability in the absence of $H_2S$.

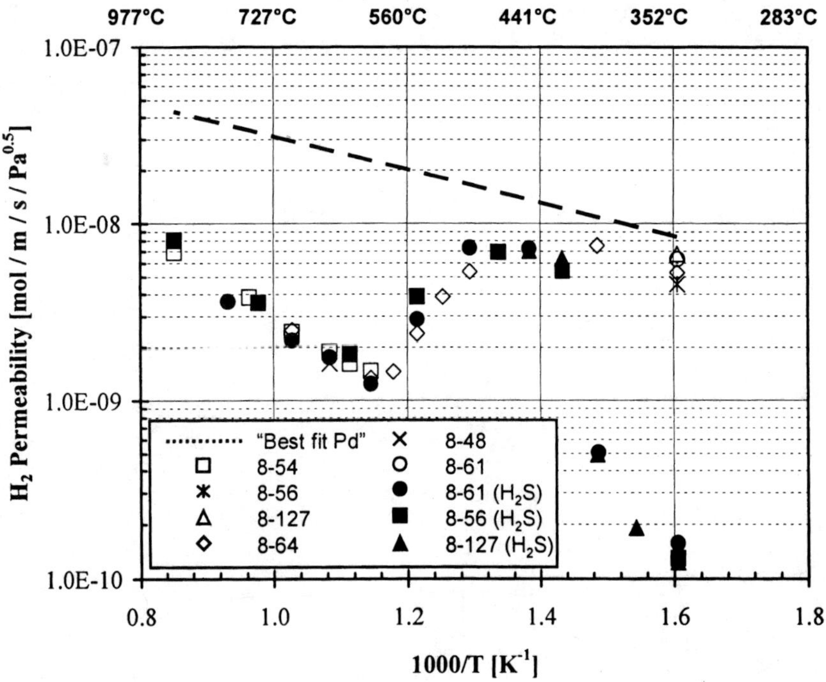


***Figure 4.*** *Steady-state hydrogen permeability results for the 1000 μm thick 60wt%Pd membrane as a function of inverse temperature under $H_2$ and 1000 ppm $H_2S/H_2$. (Open symbols signify $H_2$ and filled symbols signify 1000 ppm $H_2S/H_2$. Hyphenated numbers in figure key are the membrane identification numbers.)*

Figure 5 shows the permeability of the 53wt%Pd alloy membrane in the presence of 1000 ppm $H_2S$ over the 350 to 900°C temperature range. The permeability values again showed two distinct trends when compared to the performance in clean $H_2$. At 350°C, the permeability in the presence of $H_2S$ was two orders of magnitude lower than in $H_2$. As temperature increased, the permeability in the presence of $H_2S$ converged with that for $H_2$ until they were essentially equal at about 500°C. This range corresponds to the temperature stability range for bcc structure for this alloy composition. From this temperature, going through the transition to completely fcc in structure at 600°C and continuing up to 900°C, the permeability in the presence of $H_2S$ essentially matched the permeability in the absence of $H_2S$.

Membrane retentate-side surfaces were examined by SEM/EDS following test completion, typically 900°C. Figure 6 shows a typical membrane surface after exposure to the complete test series. The surfaces were found to be free of detectable sulfur compounds and exhibited significant grain growth. Both observations are consistent with what would be expected following the final 900°C test condition. At 900°C, under the test atmosphere composition of 1000 ppm $H_2S$, Pd and Cu sulfides are not stable[26].

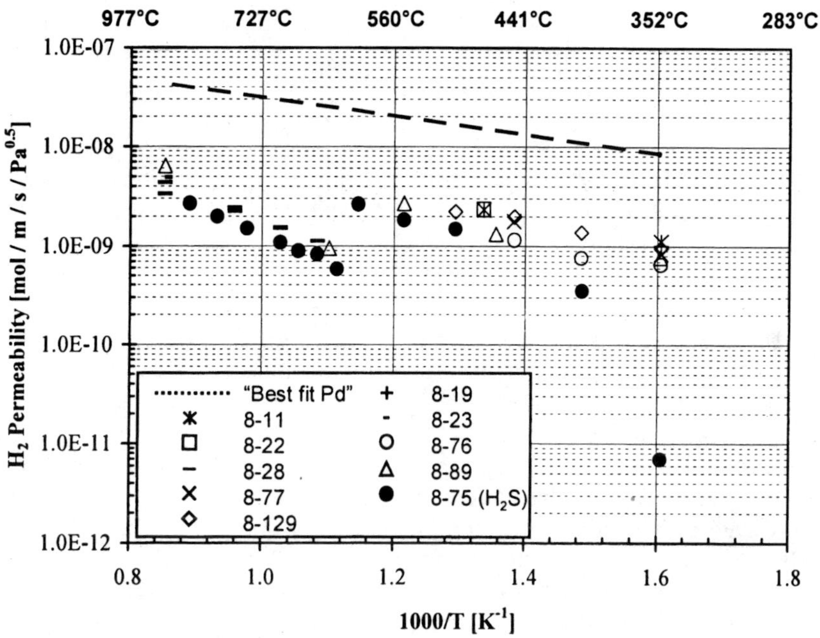


***Figure 5.*** *Steady-state hydrogen permeability results for the 1000 μm thick 53wt%Pd membrane as a function of inverse temperature under $H_2$ and 1000 ppm $H_2S/H_2$. (Open symbols signify $H_2$ and filled symbols signify 1000 ppm $H_2S/H_2$. Hyphenated numbers in figure key are the membrane identification numbers.)*

As stated previously, membranes were typically not characterized until the 1000 ppm $H_2S$, 350 to 900°C test series was completed, however, several 1000 μm membranes that failed at lower temperatures before the completion of the test series were examined. In some cases, metal sulfide growth was detected on the membrane surface by EDS. Unfortunately, for these instances, direct correlation of sulfide formation with test condition was not easily feasible since the reactor conditions were either not known (power failure) or were not able to be optimized for sample preservation.

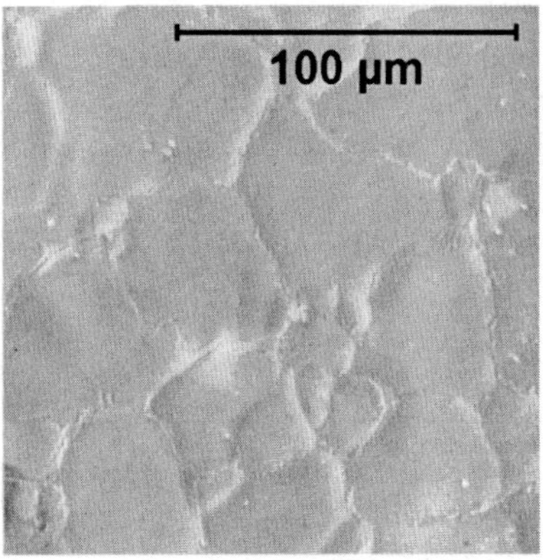


*Figure 6. SEM image showing a typical membrane surface (80wt%Pd) after hydrogen permeability testing over the 350 to 900°C range under flowing 1000 ppm $H_2S$ in $H_2$.*

The steady-state permeability results obtained for the 1000 μm thick 80, 60 and 53wt% alloys under flowing 1000 ppm $H_2S/H_2$ were compared to the transient 1000 ppm $H_2S/H_2$ permeability reported previously[21]. Nearly identical trends in permeability were found for these three alloys using these two measurement methods. This agreement reinforces the suggested correlation between the alloy crystal structure and $H_2S$ tolerance. The results of these steady-state permeability experiments indicated that when the Pd-Cu alloys had an fcc structure, $H_2S$ had little impact on permeability but when the structure was bcc, $H_2S$ had a moderate to severe impact. However, because of the extreme 1000 μm thickness of the membranes used in this test series which could potentially mask effects arising from sulfur interactions, methods were sought to enable continuous $H_2S$ exposure of 100 μm and thinner membranes.

## Steady-state $H_2S$ exposure of 100 μm foils - new mounting method

A new mounting method utilizing a Swagelok® VCR fitting was developed to enable the exposure of 100 μm thick and thinner membrane foils to flowing $H_2S$-containing test gases while avoiding the problems associated with direct brazing or welding. Initial membrane tests were carried out successfully using this method combined with a new testing protocol. First the membrane foils of interest were annealed and analyzed prior to testing to verify the crystalline structure prior to mounting in the compression fixture. The mounted membrane was heated to the selected test temperature under an inert atmosphere. Once the target temperature was achieved, the $H_2$ test gas was introduced and the baseline flux was verified. The 1000 ppm $H_2S$-containing test gas was introduced and the performance determined for a period of 120 hours. During tests, the $H_2S$ concentration exiting the reactor was monitored to verify that the bulk gas phase concentration flowing past the membrane was about 1000 ppm. Following this exposure, the inert gas flow was restarted and the system cooled as quickly as feasible. This test method allows characterization of the membrane following exposure to a specific set of conditions and partially preserves the membrane surface for the determination of surface changes. However, this method, while improved, still has limitations that must be considered during characterization such as the limited maximum cooling (quench) rate and changes in surface chemistry due to phase stability.

Figure 7 shows the flux versus time plots for 100 μm thick 100wt%Pd and 80wt%Pd membranes exposed to flowing 1000 ppm $H_2S$ for 120 hours at 350°C. For reference, the performance of each membrane in $H_2$ is also shown.

When the Pd membrane was exposed to 1000 ppm $H_2S$, an immediate flux decrease of about 25% was observed, Figure 7(a). Then over the rest of the 120 hours of exposure, the flux continued to decrease until at 120 hours it had dropped by about 70%. Preliminary examination of the exposed membrane indicated that a highly modified, very porous surface had developed as shown in Figure 8(a). EDS of the membrane cross-section indicated about an 18 μm thick layer of sulfide had grown on the membrane surface during testing, Figure 8(b).

Exposure of the 80wt%Pd alloy membrane to 1000 ppm $H_2S$ resulted in an immediate flux decrease of about 99%, Figure 7(b). Over the rest of the 120 hours of exposure, the flux decreased only slightly more. Preliminary examination of the exposed membrane again indicated a highly modified, porous appearing surface had developed although with a somewhat different appearance than that observed for Pd, Figure 8(c). Lines with a different morphology than most of the surface but the same composition were apparent. These lines probably reflect surface defects in the starting membrane such as scratches. EDS again suggested a relatively thick sulfide layer had grown on the membrane surface during testing, Figure 8(d). This layer appears to be thinner than the layer on the palladium membrane but more complex in structure as shown by the element maps of the cross-section.

## Conclusions and Future Work

Attempts at determining the effect that the continuous exposure to flowing 1000 ppm $H_2S$ had on the permeability of 100 μm thick Pd and Pd-Cu alloy foils brazed to support rings were not successful. Membrane failure was found to generally occur at about 500 to 600°C and was linked to the attack of S on the membrane-braze-support interface.

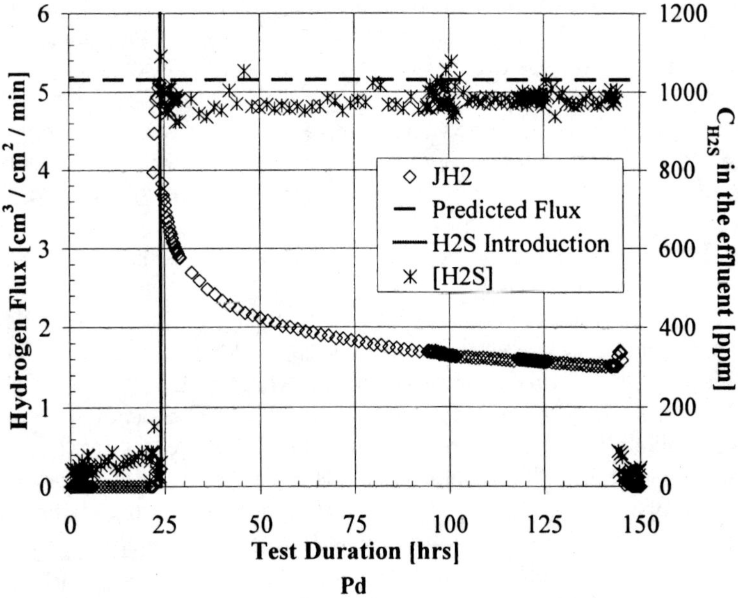


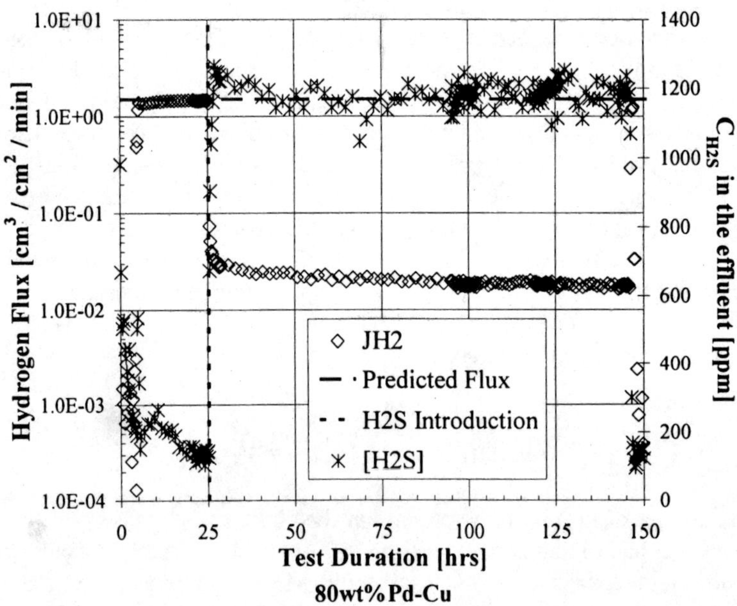


***Figure** 7. Influence of 1000 ppm $H_2S$ at $350^oC$ and ~620 kPa on the flux of pure Pd and an 80wt%Pd alloy as a function of time.*

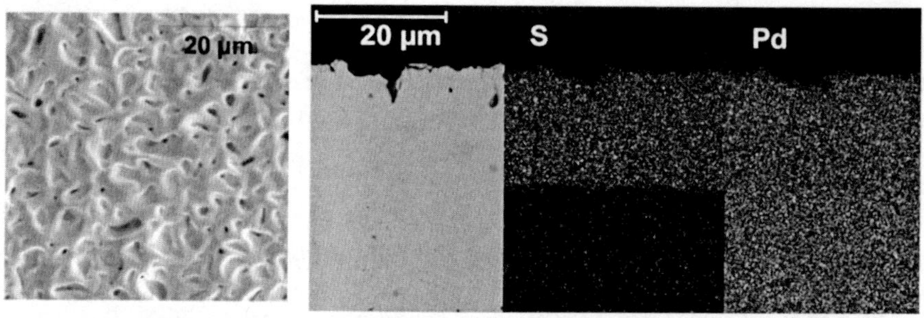


(a) Pd (b)

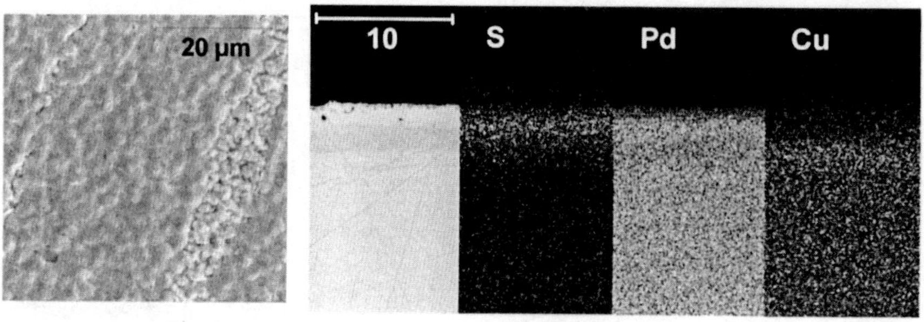


(c) 80wt%Pd- (d) 80wt%Pd-

***Figures 8****. SEM images of the Pd and 80wt%Pd membranes after 120 hrs of operation at 350°C in the presence of 1000 ppm $H_2S$ at 620 kPa. (a) Surface image of Pd. (b) EDS map of the Pd cross-section. (c) Surface image of 80wt%Pd. (d) EDS map of the 80wt%Pd cross-section.*

The permeability trends found in this study for the 1000 µm thick membranes under flowing 1000 ppm $H_2S/H_2$ were compared to transient permeability results reported previously[21]. Nearly identical trends in permeability were found for the three Pd-Cu alloys by these two measurement methods reinforcing the suggested correlation between the alloy crystal structure and $H_2S$ tolerance. The results of these steady-state permeability experiments indicated that when the Pd-Cu alloys had an fcc structure, $H_2S$ had little impact on flux but when the structure was bcc, $H_2S$ had a moderate to severe impact. However, both of these methods had limitations that probably impacted the results. The transient or batch method was limited by finite $H_2S$ availability, competition for the available $H_2S$ by other metal surfaces, relatively short test durations and a limited data set. The steady-state 1000 µm thick membrane method was potentially limited by membrane thickness which could have masked the effects resulting from sulfur interaction with the membrane surface. For example, decreases in permeability due to surface effects would be less significant and less discernible due to the diminished permeability of the thicker membrane. Additionally, since a single test membrane was exposed to a series of conditions, any surface changes resulting from $H_2S$ interaction could be lost at the next test condition.

To address these limitations, a new membrane testing protocol was introduced which enabled 100 μm and thinner membranes to be successfully tested in the presence of $H_2S$ over a long duration. In addition, an individual test membrane was exposed to only a single condition to preserve surface changes for characterization. Preliminary flux tests were conducted using 100 μm thick 100wt%Pd and 80wt%Pd membranes exposed to flowing 1000 ppm $H_2S$ at 350°C for a duration of 120 hours. These initial test results indicated that $H_2S$ had a more drastic impact on flux than was apparent through the transient or thick-membrane-steady-state results showing that the suspected limitations did influence those results. For example, the permeability decrease at 350°C of the 1000 μm thick 80wt%Pd foil exposed to flowing 1000 ppm $H_2S$ was about 20% while the decrease noted for the 100 μm thick foil was about 90%, both relative to their baseline $H_2$ performances. Characterization of the 100wt%Pd and 80wt%Pd membranes after testing revealed a significant surface sulfide growth as well as major surface morphological changes. These results suggest that, under these specific conditions, very thin membranes or coatings composed of Pd or 80wt%Pd-Cu could have a very limited useful life and could thus be impractical.

Pd-Cu alloy studies are continuing using this new approach and the results are expected to provide valuable insights into the performance of these alloys under an extensive range of conditions, but many questions remain to be answered. Future investigations include studying the effect of $H_2S$ concentration, the effect of other gas species, the structure and chemistry of the alloy surfaces and Pd-Cu alloy modifications to improve S tolerance.

## Acknowledgement and Disclaimer

The NETL HMT units were operated and maintained by Michael Ciocco, Ron Hirsh, Paul Dieter, Bill Brown, George Schlata, and Jack Thoms of Parsons. Reference in this paper to any specific commercial product, process, or service is to facilitate understanding and does not imply its endorsement or favoring by the U.S. DOE.

# References


1. Hydrogen Posture Plan: An Integrated Research, Development and Demonstration Plan. *US Department of Energy*, *www.hydrogen.energy.gov* **2004**.
2. FutureGen: Integrated Hydrogen, Electric Power Production and Carbond Sequestration Initiative. *US DOE Office of Fossil Energy*, *www.fossil.energy.gov/programs/powersystmes/futuregen/index.html* **2004**.
3. Edlund, D. J.; Pledger, W. A., Thermolysis of hydrogen sulfide in a metal-membrane reactor. *Journal of Membrane Science* **1993**, *77, (2-3)*, 255.
4. McKinley, D. L.; Nitro, W. Metal Alloy for Hydrogen Separation and Purification. 1967.
5. Edlund, D.; Friesen, D.; Johnson, B.; Pledger, W., Hydrogen-permeable metal membranes for high-temperature gas separations. *Gas Separation & Purification* **1994**, *8, (3)*, 131.
6. Kajiwara, M.; Uemiya, S.; Kojima, T., Stability and hydrogen permeation behavior of supported platinum membranes in presence of hydrogen sulfide. *International Journal of Hydrogen Energy* **1999**, *24*, 839.
7. Kulprathipanja, A.; Alptekin, G. O.; Falconer, J. L.; Way, J. D., Pd and Pd-Cu membranes: inhibition of $H_2$ permeation be $H_2S$. *Journal of Membrane Science* **2005**, *254*, 49-62.
8. Hurlbert, R. C.; Konecny, J. O., Diffusion of Hydrogen through Palladium. *The Journal of Chemical Physics* **1961**, *34, (2)*, 655.
9. Lalauze, R.; Gillard, P.; Pijolat, C., Hydrogen permeatioin through a thin film of palladium: influence of surface impurities. *Sensors and Actuators* **1988**, *14, (3)*, 243.
10. Darling, A. S., Hydrogen separtaion by diffusion through palladium alloy membranes. *Symposium on the less common means of separation, Institution of chemical engineers* **1963**.
11. Ali, J. K.; Newson, E. J.; Rippin, D. W. T., Deactivation and regeneration of Pd-Ag membranes for dehydrogenation reactions. *Journal of Membrane Science* **1994**, *89, (1-2)*, 171.
12. Edlund, D., A Membrane Reactor for $H_2S$ Decomposition. In *Advanced Coal-Fired Power Systems*, Bend Research, Inc.: Bend, 1996.
13. Bryden, K. J.; Ying, J. Y., Nanostructured palladium-iron membranes for hydrogen separation and membrane hydrogenation reactions. *Journal of Membrane Science* **2002**, *203, (1-2)*, 29.
14. Kusakabe, K.; Yokoyama, S.; Morooka, S.; Hayashi, J. i.; Nagata, H., Development of supported thin palladium membrane and application to enhancement of propane aromatization on Ga-silicate catalyst. *Chemical Engineering Science* **1996**, *51, (11)*, 3027.

15. Edlund, D. P., W., Catalytic Platinum-based membrane reactor for removal of H2S from natural gas streams. *Journal of Membrane Science* **1994,** 94, 111-119.
16. Edlund, D. J.; McCarthy, J., The relationship between intermetallic diffusion and flux decline in composite-metal membranes: implications for achieving long membrane lifetime. *Journal of Membrane Science* **1995,** 107, (1-2), 147.
17. Kamakoti, P.; Morreale, B. D.; Ciocco, M. V.; Howard, B. H.; Killmeyer, R. P.; Cugini, A. V.; Sholl, D. S., Prediction of Hydrogen Flux Through Sulfur-Tolerant Binary Alloy Membranes. *Science* **2005,** 307, (5709), 569-573.
18. Alfonso, D.; Cugini, A.; Sholl, D., *Surface Science* **2003,** 546, 12.
19. Alfonso, D.; Cugini, A.; Sorescu, D., *Catalysis Today* **2005,** 99, 315.
20. Howard, B. H.; Killmeyer, R. P.; Rothenberger, K. S.; Cugini, A. V.; Morreale, B. D.; Enick, R. M.; Bustamante, F., Hydrogen permeance of palladium-copper alloy membranes over a wide range of temperatures and pressures. *Journal of Membrane Science* **2004,** 241, (2), 207.
21. Morreale, B. D.; Ciocco, M. V.; Howard, B. H.; Killmeyer, R. P.; Cugini, A. V.; Enick, R. M., Effect of hydrogen-sulfide on the hydrogen permeance of palladium-copper alloys at elevated temperatures. *Journal of Membrane Science* **2004,** 241, (2), 219.
22. Morreale, B. D.; Ciocco, M. V.; Enick, R. M.; Morsi, B. I.; Howard, B. H.; Cugini, A. V.; Rothenberger, K. S., The permeability of hydrogen in bulk palladium at elevated temperatures and pressures. *Journal of Membrane Science* **2003,** 212, (1-2), 87.
23. Rothenberger, K. S.; Howard, B. H.; Killmeyer, R. P.; Cugini, A. V.; Enick, R. M.; Bustamante, F.; Ciocco, M. V.; Morreale, B. D.; Buxbaum, R. E., Evaluation of tantalum-based materials for hydrogen separation at elevated temperatures and pressures. *Journal of Membrane Science* **2003,** 218, (1-2), 19.
24. Okamoto, H., Phase Diagrams for binary alloys. **2000.**
25. McKinley, D. L. Method for Hydrogen Separation and Purification. 1969.
26. Barin, I.; Sauert, F.; Schultze-Rhonhof, E.; Sheng, W., Thermodynamic properties of inorganic substances. **1993.**

# Upgrading

## Chapter 12

# Removal of Nitrogen from Liquid Fuels by Supercritical Fluid

**Olubunmi M. Ogunsola**

**Tryby Energy Minerals & Environmental Corporation (TEMEC), P.O. Box 1072, Woodbridge, VA 22195**

Liquid products derived from fossil fuels such as oil shale, coal and tar sands are required to be upgraded to be suitable as substitutes for petroleum. Upgrading is mainly done to enhance the H/C ratio and reduce the nitrogen and sulfur heterocyclic compounds which are present in large quantities in these syncrudes. The coneventional way of achieving this is by hydrotreating, i.e. by reacting the syncrude with hydrogen at high pressure in the presence of suitable catalysts. Sulfur and nitrogen are removed as $H_2S$ and $NH_3$. However, the quantity of hydrogen required can drive the cost of this process up. An innovative method of removing nitrogen compounds from syncrude is by supercritical water. The separation process is milder than hydrodenitrogenation which is the conventional nitrogen removal method. The results of previous studies aimed at supercritical water denitrogenation under different operating conditions are reviewed. The denitrogenation of model compounds gives an insight into the mechanism of denitrogenation process.

## Resources of oil shale and tar sand

The U.S. resources of oil shale and tar sands are estimated at about 2 trillion barrels and about 76 billion barrels, respectively. These huge unconventional oil resources are currently undeveloped (*1,2*). To develop these unconventional domestic energy resources several constraints, which include the presence of nitrogen and sulfur, which can lead to increased emission of nitrogen oxides and sulfur dioxide respectively, need to be addressed.

In order to obtain high value transportation fuels that will meet the new, more stringent environmental requirements, the sulfur and nitrogen contents have to be considerably reduced. Current conventional method of nitrogen and sulfur removal by hydro-treating is very expensive due to huge demand of expensive hydrogen for their effective removal. Supercritical fluids have been successfully used to break the carbon-nitrogen and carbon-sulfur bonds in model compounds by a number of researchers, including Ogunsola and Berkowitz (*3*) and Ogunsola (*4,5*). Since the nitrogen and sulfur contained in oil shale and tar sands are similar to these model compounds, supercritical fluids are therefore good candidates for removing nitrogen from shale oil and sulfur from tar sands-derived crude.

This paper discusses previous studies aimed at determining the effectiveness of removing nitrogen from shale oil by supercritical water (SCW).

## Distribution of nitrogen in oil shale and tar sand

In the past, several attempts have been made to recover these hydrocarbons, but there has been limited success due to several constraints. The constraints in the development of these domestic tar sands and other unconventional resources include geological challenges, high costs of extraction, production and upgrading, and environmental concerns. The most significant drawback to the use of finished transportation fuels produced from these unconventional resources is their high potential for NOx (nitrogen oxides) emission as a result of their high nitrogen content. For example, raw shale oil contains about twice as much nitrogen as that typically found in petroleum crude. This property significantly impacts the refining/upgrading of these synthetic crude oil. Most refining operations that utilize catalysts cannot function at such high nitrogen levels since these catalysts are readily poisoned by nitrogen compounds. Other problems caused by high amounts of nitrogen in syncrude are product instability and limited storage life, contamination during transportation in common crude oil pipelines, mutagenic properties and excessive NOx emission during combustion of the fuels. These problems strongly support the need to remove nitrogen from crude shale oil.

Major nitrogen compounds in petroleum crude oil have been characterized mainly into four classes: pyridines, diaza compounds, carbazoles, and amides

(*6*). It was also reported that pyridines and pyrroles were the two major types of nitrogen compounds in shale oil (*7*). Also, Brown and co-workers (*8*) identified alkyl-pyridines, cycloalkanopyridines, alkylanilines, quinolines, tetrahydroquinolines, and tetrahydroisoquinolines in hydrocracked shale oil naphtha (*8*).

Overall, shale oil contains about 61% nonhydrocarbons. Studies have shown that there are two major nitrogen compounds that are found in shale oil: pyridines and pyrroles (*6*). Other compounds that are often found include quinolines, indoles, and pyridine derivatives. Table 1 shows representative nitrogen-containing compounds in petroleum, shale oil and coal derived liquids, while Figure 1 shows the model structure of oil shale with nitrogen widely distributed. In addition, analysis has also discovered nitrogen-containing aromatics such as azarenes, azadibenzothiophenes, and cyclic amides. It is safe to say that nitrogen structures found in oil shale are primarily aromatic, with negligible quantities of saturated nitrogen compounds. It should also be noted that the nitrogen distribution of different compound types in shale oil derived from kerogen and bitumen is not necessarily the same (*9*).

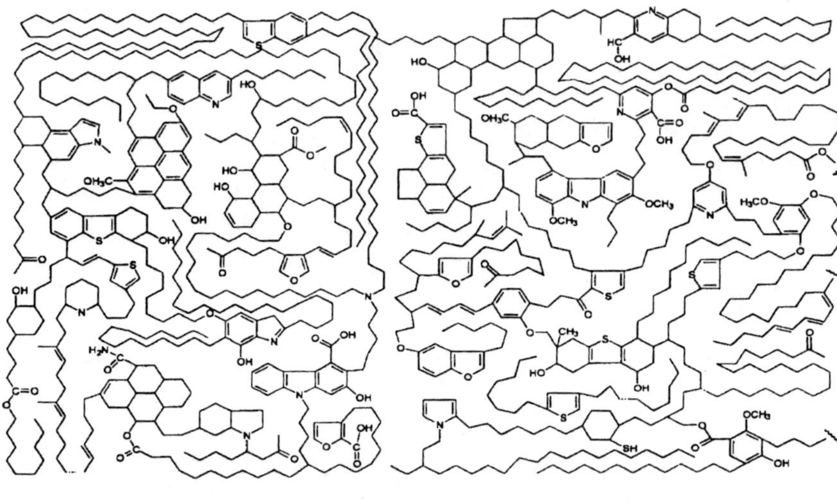

*Figure 1. Sctructural model of oil shale (10)*

## Impact of nitrogen on syncrude storage stability

The presence of alkylated heterocyclic nitrogen compounds in processed shale oil can cause sedimentation. This especially becomes problematic in

**Table1. Representative Nitrogen-Containing Compounds in Petroleum, Oil Shale and Coal-Derived Liquids (*10*)**

| *Compounds* | | |
|---|---|---|
| *Nonheterocyclic compounds* | | |
| Aniline 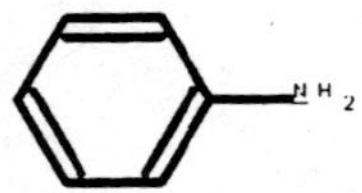  | Pentylamine   | |
| *Nonbasic heterocyclic compounds* | | |
| Pyrrolc   | Indolc 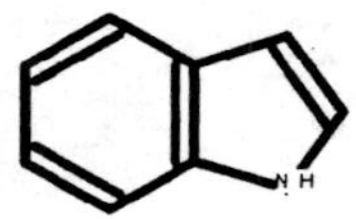  | Carbazolc 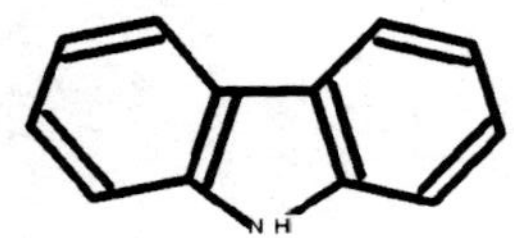  |
| *Basic heterocyclic compounds* | | |
| Pyridine   | Quinoline 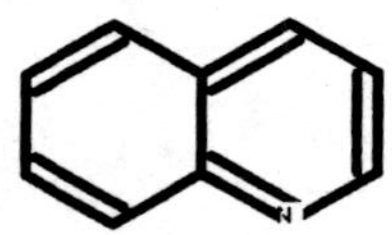  | Indoline 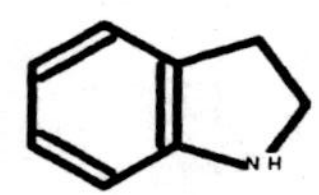  |

| Acridine | Benz(a) acridine | Benz(c) acridine |
| --- | --- | --- |
| 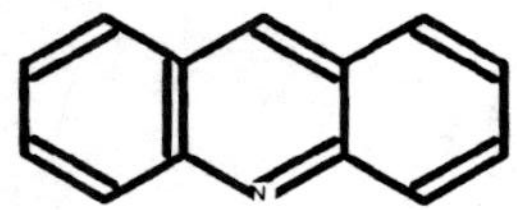 | 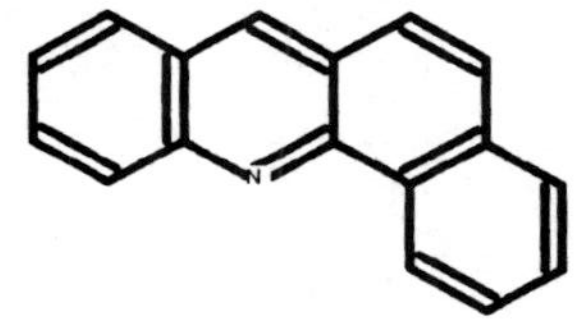 | 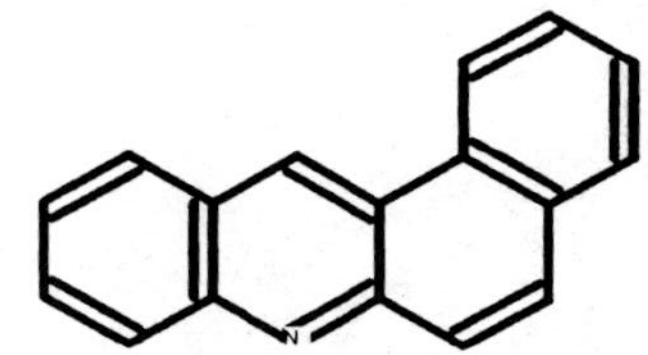 |

compounds where the alkyl groups are adjacent to nitrogen. Fortunately, basic nitrogen compounds or nonalkylated heterocycles typically form only modest amounts of sediment. However, in the presence of reactive nitrogen compounds they can promote sedimentation. In addition, sedimentation from pyrrolic compounds can occur from oligomerization through adjacent methyl groups (*11*). The possibility of sedimentation is an important factor to consider when obtaining shale oil. Not only does sedimentation affect the storage stability of the shale oil, but it can also increase the cost and difficulty in processing and upgrading it. The storage stability of crude shale oil is very important considering the fact that retorting and upgrading are typically done in different locations. Also, a significant amount of time typically passes before the shale oil is processed.

## Consequences of nitrogen in syncrude processing

### Effect on catalyst

Nitrogen compounds are poisonous to catalysts and usually interfere with catalytic reactions such as catalytic cracking. The reason for this lies in the fact that nitrogen compounds are very basic and preferentially absorb on the acidic catalyst surface (*12*).

### Environmental

Having nitrogen-containing compounds in transportation fuels is detrimental to our environment. When a fuel is burned at very high temperatures, NOx are formed. These compounds have been associated with not only acid rain, but also the formation of fine particulate matter and ground level ozone (smog).

## Nitrogen removal methods/processes

### Hydrodenitrogenation

One of the most extensively used methods to remove nitrogen from oil products is hydrodenitrogenation. In this hydrotreating process, liquid oil products are exposed to hydrogen in the presence of suitable catalysts.

Hydrotreating is the largest application of industrial catalysts based on the amount of material processed each year, and hydrotreating catalysts make up the third largest business based upon the amount of catalyst sold each year (*13*). During the process to remove nitrogen-containing molecules, hydrogenation of the aromatic rings must occur first. Due to the aromaticity of the nitrogen-containing ring in polycyclic aromatics, the nitrogen atom can only be removed after hydrogenation of this ring which generally requires relatively high pressures of hydrogen. After this step, the C-N bond is cleaved as a direct result of $NH_3$ elimination (*13*). Studies have proven that in this type of reaction, both hydrogen consumption and the overall rate of denitrogenation increase with pressure (*14*).

Among the models which are representative of the nitrogen impurities of petroleum, quinoline is one of the most simple, the decomposition of which involves all three reactions that are likely to occur during the hydrodenitrogenation of such impurities (the hydrogenation of nitrogen aromatic rings, the hydrogenation of benzene rings and carbon-nitrogen bond cleavages). Moreover, 1,2,3,4-tetrahydroquinoline makes it possible to compare at once the activities of the catalysts in carbon-nitrogen bond cleavage and in hydrogenation of benzene rings. The reaction scheme for quinoline hydrodenitrogenation proposed by Satterfield et. al (*15*) is shown in Figure 2.

## Ion-exchange

Another process that has been found to be effective at removing nitrogen from transportation fuels utilizes ion exchange resins. A few ion exchange resins, or adsorbents, that have been used experimentally include Amberlyst-15, Amberlyst XN-1010, and Amberlite XE-397. All three of these resins exist as sulfonic acid on a styrene-divinyl-benzene backbone. When tested for nitrogen removal, they exhibited a significantly larger capacity for adsorption than other "inorganic" adsorbents. Other resins containing carboxylic acid functionality on a polyacrylic backbone exhibited no effective adsorption at all (*16*).

Although the aforementioned resins were effective at removing nitrogen from shale-derived jet fuels in particular, they did not do so within the same capacity. For example, the integrated nitrogen adsorption capacity of XE-397 was about 2.5 times more effective than that of A-15 (*16*). In addition, XE-397 had utilization capacity greater than 100% in comparison to its theoretical adsorption capacity (*16*). Though all three acids had similar acid functionality, their different pore size distributions seemed to have a significant effect on their adsorption capacities (*16*). As a result, it should not be surprising that XE-397 had the greatest adsorption capacity considering it also had the biggest mean

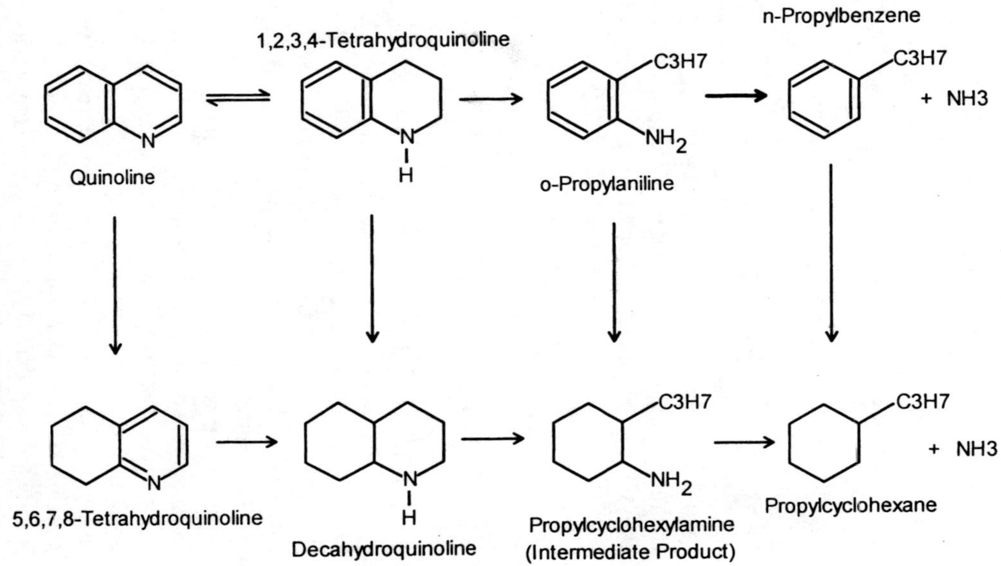


*Figure 2. Reaction athways in quinoline hydrodenitrogenation (15 )*

pore size. The correlation between pore size and adsorption capacity is of no consequence considering that ion-exchange processes are assumed to be controlled by intraparticle diffusion (*17*). Thus, not only is it necessary to understand the chemical aspects of an ion-exchange process, understanding the physical aspects of the process are necessary as well.

### Adsorption

Another way to separate nitrogen compounds from hydrotreated shale oil products is by adsorption chromatography. Experiments have been conducted where basic and neutral alumina have been used to separate shale oil into specific nitrogen-type fractions. Both a pumped-flow and gravity-flow method were executed. However, it was determined that the pumped-flow procedure was better since it enabled more control over experimental parameters such as flow rate. This separation scheme could be very useful because it has been applied to shale oil products produced under different hydrotreating conditions (*18,19*).

### Solvent extraction

It has been found through experimentation that ionic liquids can be used to extract nitrogen from transportation fuels. In one study, 1-alkyl-3-methylimidazolium (AMIM) tetrafluoroborate and hexafluorophosphate were used for the purpose of denitrogenation (*20*). The liquids proved to have a strong propensity for absorbing compounds with an aromatic structure. In fact, their absorption capabilities decreased as the aromatics became hindered by alkyl groups in the aromatic rings (*20*). This strong affinity for aromatic compounds is most likely related to the high polarity of the ionic liquids.

There is a strong interaction between the polarized aromatic molecules and the charged ion pairs of the ionic liquids. From these observations, one may conclude that molecules with high polarizable п-electron density prefer the molecular structure of ionic liquids (*20*).

### Supercritical fluid (water) for denitrogenation

All the above methods have their set backs. For example hydrodenitrogenation requires a large quantity of hydrogen which can be expensive. Similarly solvent extraction and adsorption processes are plagued with high cost for materials. Besides, products from solvent extraction process can be contaminated with residues left from the solvents. However, supercritical water extraction is a novel process that is relatively cheaper and more effective. It uses cheap water as a source of much needed hydrogen. This and other advantages are discussed later in this paper.

## Fundamentals of supercritical fluid

Under supercritical conditions, fluids behave like dense gases with high diffusivities (*21*), resulting in high mass transfer. This property allows supercritical fluids to dissolve non-volatile compounds more easily. It has been reported that the reaction during supercritical fluid extraction varies from gas-like to liquid-like as the solvent density increases (*22*). It is therefore expected to observe two different reaction pathways, depending on density. Some other unique characteristics of supercritical fluids include minimal chemical degradation of the extracts and simple recovery of these extracts.

Water is an example of a liquid that works well for supercritical fluid extraction/upgrading. Liquid water at standard conditions can be an excellent solvent for many compounds and electrolytes because of its high dielectric constant, but it is negligibly miscible with hydrocarbons and gases. However, in the supercritical region, water becomes completely miscible with hydrocarbons and gases. In this region, there are no phase boundaries and many reactions may proceed without many mass transfer limitations (*23*).

One of the most important properties of water is its dielectric constant. Not only can this parameter be varied by temperature and pressure, it can also greatly influence the rate of chemical reactions. The ion product of water ($K_w$) also has a strong influence on chemical reactivity. Since the ion product can be significantly higher in the supercritical region, water may play the role as an acid or base catalyst. Acid or base catalyzed reactions in water exhibit non-Arrhenius behavior at high temperatures and pressures which can drastically affect reaction rates (*23*). Both the dielectric constant and ion product of supercritical water are some of the properties that are responsible for its solvent power. Other particular advantages of water as a solvent for supercritical fluid extraction/upgrading are: greatly reduced further contamination of extracts, non-toxicity, low cost, and high extractability of heavy hydrocarbons.

## Supercritical fluid processing of syncrude using sand bath

In order to evaluate the potential of denitrogenating syncrude, some experiments were conducted in micro bomb tubing reactors constructed with swagelock fittings. In each run, the micro reactor was loaded with 2 g of sample and water. After loading and closing the reactor, nitrogen was used to purge the reactor before pressurizing the system to the cold pressure that will generate the hot reaction pressure. The reactor was then immersed in a preheated fluidized sand bath and the time of immersion recorded as the start of the experiment. Iron Oxide ($Fe_2O_3$) catalysts was added to some reactions to determine the catalytic effect. The reactions were carried out at 300°C and 400°C, at 14 and 22MPa, for various durations ranging from 1 hour to 24hours. Details of the experimental procedure are discussed elsewhere (*4*).

Quinoline was chosen for the denitrogenation experiment with the assumption that its behavior is representative of many other nitrogen heterocyclic compounds present in syncrude from tar sand and oil shale.

The results of supercritical water reaction of quinoline in the presence of $Fe_2O_3$ conducted by Ogunsola (*4*) allowed her to develop a reaction sequence depicted in Figure 3. Generation of aniline type products has been regarded by Houser et al (*24*) as due to pronounced preference for 1-2 bond rupture. The presence of naphthalene 2- methyl also suggests that naphthalene is an intermediate product that may arise from methyl-indane interaction.

Formation of hydroxyl quinoline must have been initiated by ionic reactions that involved $H^+$ and $OH^-$ ions. Hydrogenation by $H_2$ enables the heterocyclic ring to be saturated, and this can be followed by hydrogenolysis of C-N bonds that first open the hetero-ring and then convert the resultant aliphatic and aromatic amine intermediates to hydrocarbons and ammonia (*24*). Also direct attack of $OH^-$ on quinoline may have led to formation of hydroxyl quinoline. It has been suggested that both ionic reaction and free radical capping are possible under supercritical conditions (*25*).

The formation of 2,2-biquinoline from pyrolysis of quinoline suggest that the main reaction mode is thermal cracking and subsequent random polymerization to heavier molecules.

Aliphatic amines such as pyridine are very reactive and are not substantial constituents in the original feeds, but they are formed as intermediates during hydrodenitrogenation (HDN) of cyclic nitrogen compounds. For example, the HDN of pyridine proceeds via saturation of the heterocyclic ring followed by ring opening to *n*-pentylamine and subsequent removal of the nitrogen by deamination (*26*). Considerable savings in hydrogen and a better hydrocarbon product would be obtained if the catalytic activity for C-N bond cleavage could be enhanced. Therefore, it is important to examine the last step corresponding to the C-N bond cleavage of the primary amine. The mechanism most often mentioned for amine removal reactions is the Hoffmann degradation (*27, 28*).

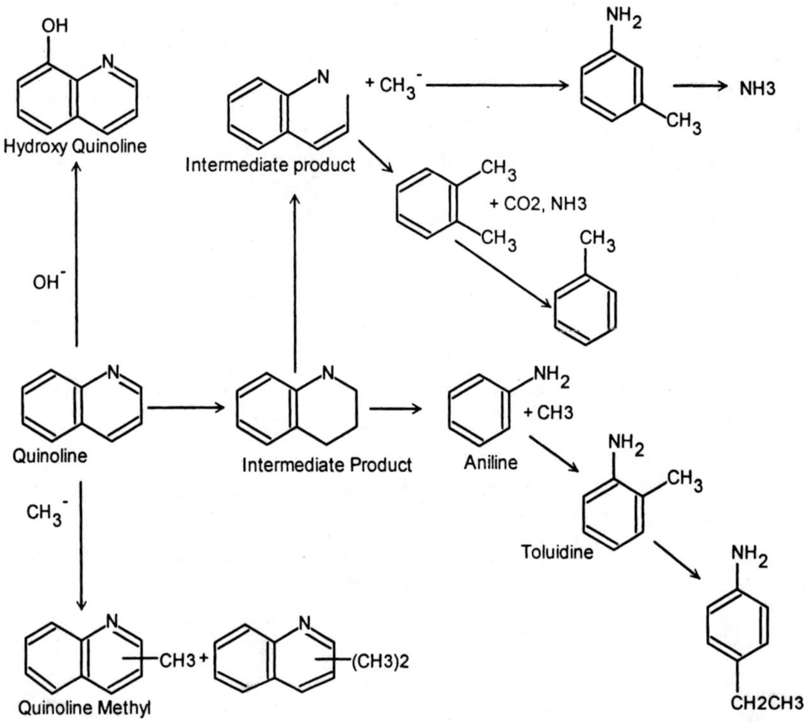


*Figure 3. Proposed reaction sequence in quinoline denitrogenation by SCW(4)*

This mechanism usually requires that the leaving nitrogen be quaternized, and the degradation is known to occur only with saturated hydrocarbons. The N removal step is either a β-elimination, involving a hydrogen of the carbon in the β position with respect to the nitrogen atom (*27*), or a nucleophilic substitution (*27*). Under the experimental conditions required for HDN, the olefinic compounds formed can be readily hydrogenated.

In summary, it has been demonstrated from various previous studies that supercritical water can be used to upgrade and denitrogenate syncrude derived from unconventional hydrocarbon such as tar sand and oil shale. The presence of $Fe_2O_3$ catalyst enhanced the denitrogenation process. A perusal of the literature reveals that SCW appears to show more promise in removing nitrogen from nitrogen-containing fuels than conventional denitrogenation methods.

## References

1. Dyni, J.R. "Geology and Resources of some World Oil Shale Deposits". *Oil Shale,* **2003**, *20*, 193-252.
2. Congress of United States, Office of Technology assessment, an Assessment of oil Shale Technologies, June **1980** (OTA Report No. OTA-M-118).
3. Ogunsola, O. M.; Berkowitz, N. Fuel **1995**, *74(10)*, 1485-1490.
4. Ogunsola, O. M. PhD Thesis, The University of Alberta, Edmonton, Canada, **1991**.
5. Ogunsola, O. M. J.Hazardous Mat. **2000**, *B74*, 187-195.
6. Shue, Feng-Fang and Yen, The Fu. Anal. Chem **1981**, *53*, 2081-2084.
7. Poulson. R.E., Frost' C.M.; Jensen, H.B. In "Shale Oil, Tar Sands and related Fuel Sources"; Yen, T.F.' Ed. American Chemical Society: Washington, DC, **1976**; Adv. Chem. Ser. pp 1-10.
8. Brown, D.; Earnshaw, D.G.; McDonald, F.R.; Jensen, H.B. Anal. Chem. **1970**, *42*, 146.
9. Mitra-Kirtley, S.; Mullins, O.C.; Branthaver, J.F.; Cramer, S.P. Energy and Fuels **1993**, *7*, 1128-1134.
10. Qu,Lianglong. PhD dissertation, Swiss Federal Institute of Technology Zurich, **2002**.
11. Meler, P. F., Hankinson, R.W. D. K. Petree, N. K. Phillips, Parrish, W. R. Allred, G. C.Ind. Eng. Chem. Prod. Res. Dev **1986**, *25*, 355-360.
12. Giola, F. Murena, F. J. Hazardous Materials **1998**, *57*, 177-192.
13. Prins, R.; Jian, R. M.; Fleschsenhar, M. Polyhedron **1997**, *16*, 3235-3246
14. Townsend, S.H. et al., Ind. Chem. Res **1988**, *27*, 143.
15. Satterfield, C.N, M. Nodell, R.A. Hites, Declerck C.J. Ind. Eng. Chem. Process Des. Dev. **1978**, *17(2)*,141.
16. Marcelin, G.; Cronauer, D.C.; Vogel, R.F.; Prudlch, M.E.; Solash, J. Ind. Eng. Chem. Process Des. Dev **1986**, *25*, 747-756.
17. Prudlch, M.E., Cronauer, D.C., Vogel, R.F., Solash, J. Ind. Eng. Chem. Process Des. Dev. **1986**, *25*, 742-746.
18. Ford, C.D.; Holmes, S.A.; Thompson, L.F.; Latham,D.R. Anal. Chem. **1981**, *53*, 831-836.
19. Smith, J. R., C.R. Smith and G.U. Dinneen, Analytical chemistry **1952**, *22(7)*, 867.
20. Zhang, S.; Zhang, Q.; Zhang, Z.C. Ind. Eng. Chem. Res. **2004**, *43*, 614-622.
21. McHugh, M.A. in Recent Development in Separation Science Vlo.iX, li, N.N. & Calo, J.m. (Eds) CRC Prsee, Boca Raton, FL, **1985**.
22. Whitehead, J.C.; William, D.F. J. Inst. Fuel **1975**, *48*, 182.

23. Williams, D.F. Eng. Sci. **1981,** *36 (11)*, 1769.
24. Houser, T.J. et al. Fuel **1986**, *65*, 827.
25. Penniger, J.M. L. Fuel **1988**, *67*, 490.
26. Schwartz, V., PhD thesis, Virginia Tech, **2000**.
27. Zhao, Y. PhD dissertation, Swiss Federal Institute of Technology Zurich, **2004**.
28. Prins, R.; Zhao, Y.; Sivasankar, N.; Kukula, P. J. of catalysis **2005**, *234*, 509-512.

# Chapter 13

# Application of Supercritical Fluids to Solid Acid Catalyst Alkylation and Regeneration

**Lucia M. Petkovic, Daniel M. Ginosar*, David N. Thompson, and Kyle C. Burch**

**Chemistry Department, Idaho National Laboratory, P.O. Box 1625, MS 2208, Idaho Falls, ID 83415–2208**

Supercritical fluid (SCF) regeneration is a promising alternative method for regenerating solid catalysts deactivated by carbonaceous deposits. The unique solvent and transport properties of SCFs such as solvent strength similar to liquids and transport properties similar to gases make them highly suitable for extraction of fouling materials from porous heterogeneous catalysts. A brief review of the research work performed at the Idaho National Laboratory (INL) on the application of supercritical fluids to both isobutane/butene alkylation reaction and solid acid catalyst regeneration is presented in this contribution.

Alkylation processes are utilized in the petroleum refining industry to produce a low vapor pressure, high-octane gasoline blend stock through the reaction of high vapor pressure isoalkanes and alkenes. The product ("alkylate") is desired not only because of its high octane rating, blending properties, and environmental advantages, but also because it is considered to be an essential factor in meeting the growing demand of reformulated gasoline in the US (*1*).

Industrial processes use concentrated mineral acids to catalyze the alkylation reaction. This poses serious safety and environmental risks arising from the transport and storage of the concentrated liquid acids and the need to dispose of acid-oil sludges.

Solid acid catalysts could be used as a safer and more environmentally acceptable alternative to liquid acids, however there are significant technical barriers to overcome before this can be realized. A major limitation to the use of the solid acid catalysts is their deactivation due to the formation and deposition of carbonaceous deposits. Current industrial methods to regenerate solid acid catalysts involve oxidation or hydrogenation. However, oxidation may alter catalyst structure; hydrogenation requires the presence of noble metals on the catalyst and a hydrogen supply infrastructure. All these factors significantly increase capital and operating costs. Many of the technical challenges surrounding isoparaffin-olefin alkylation processes (*2*) and the application of solid acid catalysts (*3*) have been reviewed.

Supercritical fluid regeneration is a promising alternative method for regenerating solid catalysts deactivated by carbonaceous deposits. The unique solvent and transport properties of SCFs such as solvent strength similar to liquids and transport properties similar to gases make them highly suitable for extraction of fouling materials from porous heterogeneous catalysts. A review on the application of supercritical fluids in heterogeneous catalysis is available in the literature (*4*).

In this article, a review of the research work performed at the Idaho National Laboratory (INL) on the application of supercritical fluids to both isobutane/butene alkylation reaction and solid acid catalyst regeneration is presented. The first section summarizes the exploration of isobutane/trans-2-butene alkylation over six solid acid catalysts in the liquid (L), modified liquid (ML), near-critical liquid (NC-L), and supercritical (SC) regions through the addition of a cosolvent to the reaction feed mixture. The second section includes the work on the application of different fluids and fluid phases to regenerate a completely deactivated acidic ultrastable Y (USY) zeolite catalyst and the particular case of application of SC isobutane. The third section includes the research performed to explore two criteria to end the alkylation reaction and initiate SCF regeneration on a partially deactivated USY zeolite catalyst and the automated reaction/SCF regeneration application utilizing both synthetic feed and commercial refinery isoparaffin/olefin blends. The fourth section investigates the chemistry of the SC isobutane regeneration process.

# Application of SCF to the Isobutane/Butene Alkylation Reaction

Initial experiments performed at the INL compared different catalysts, fluids, and operating conditions to determine the effect of SCF on solid acid catalyst alkylation (*5*). Three sets of studies were performed: a catalyst comparison using six different catalysts (i.e., two zeolites, two sulfated metal oxides, and two Nafion catalysts) with methane as a cosolvent; an exploration of the effect of varying methane addition on alkylation using a USY zeolite catalyst; and a study of the effect of seven cosolvents (i.e., three hydrocarbons, two fluorocarbons, carbon dioxide, and sulfur hexafluoride) at L, ML, NC-L, and SCF conditions on the USY catalyst performance.

The six catalysts selected represented a variety of acidities and porosities to provide a wide range of properties. Seven different cosolvents were chosen to explore differing solvent strengths and transport properties and hence covered a wide range of densities, molecular weights, and dipole moments. All experiments were conducted in a plug-flow reactor system and the reaction was allowed to proceed until the concentration of trimethylpentanes (TMPs) in the outlet stream decreased to zero, indicating catalyst deactivation.

In general, the light hydrocarbons and trifluoromethane were the best among the different cosolvents analyzed. Catalysts that presented higher acidities (i.e., Pt/USY and USY catalysts) produced at least 4 times greater yields of $C_{5+}$ and TMP and demonstrated greater longevity than the other catalyst. However, operation under NC and SCF conditions was detrimental to catalyst longevity and product selectivity as compared to operation under liquid conditions. The application of different fluids and phases to regenerate a completely deactivated acidic USY catalyst and a more detailed analysis of the SC isobutane regeneration were undertaken next and are summarized in the following section.

# Regeneration of a Completely Deactivated USY Catalyst

## Application of Different Fluids and Fluid Phases

Activity recovery of a completely deactivated USY zeolite catalyst that initially was deactivated during the liquid phase isobutane/butene alkylation reaction was examined in a continuous plug-flow regeneration system employing a series of $C_3$-$C_5$ alkanes (*6*). In addition, helium was also utilized as a regeneration fluid to analyze the effects of volatility on catalyst activity recovery. Regeneration experiments were performed by running the reaction mixture over fresh catalyst until it was completely deactivated, flowing a regeneration fluid over the catalyst bed at specific conditions, and then rerunning the reaction mixture over the regenerated catalyst bed under conditions identical to the first step. Regeneration effectiveness was determined by dividing product yield after regeneration by the yield obtained using the fresh catalyst.

To determine the effectiveness of the different fluids on catalyst reactivation, temperature and pressure were kept constant at 111 bar and 180 °C. Under these conditions, helium was in the gas phase, propane, n-butane and isobutane were supercritical, and n-pentane and isopentane were NC liquids. The results of regeneration effectiveness (i.e., yield after regeneration with respect of yield from the fresh catalyst) are shown on Table I. For example, SC isobutane and propane recovered as much as 82% and just 24% of the fresh catalyst TMP activity, respectively. Isobutane clearly provided the greatest catalyst regeneration. The increased ability of the isobutane and isopentane to facilitate hydride transfer reactions with adsorbed deactivating high-molecular weight species was proposed to play a significant role in attaining high levels of catalyst activity recovery. This effect was confirmed when analyzing the effect of fluid phase on catalyst activity recovery employing the $C_4$ and $C_5$ alkanes at both NC liquid and SCF conditions at 111 bar. For these experiments, the temperature was adjusted to obtain the desired fluid phase. The results in **Error! Reference source not found.** show that effect of phase and the $C_4$ and $C_5$ isomers having a tertiary carbon atom in their molecule were more effective to recover catalyst activity than the linear alkanes both at NC-liquid and SCF conditions. This result confirmed that facile hydride transfer between adsorbed fouling compounds and the reactivating fluid may play a role in promoting catalyst activity recovery in addition to cracking, volatilization, and extraction. Additionally, these results confirm that the SCF phase was most effective at restoring initial catalyst activity. A more detailed study of the SC isobutane regeneration was carried out next as outlined in the following section.

**Table I. Regeneration Effectiveness for All Regeneration Fluids**

| | *helium* | *propane* | *n-butane* | *isobutane* | *n-pentane* | *isopentane* |
|---|---|---|---|---|---|---|
| | | | *(% activity recovery)* | | | |
| $C_{5+}$ | 34 | 28 | 57 | 75 | 27 | 66 |
| $C_8$ | 37 | 30 | 59 | 78 | 30 | 63 |
| TMP | 33 | 24 | 51 | 82 | 19 | 59 |

NOTE: Table based on Figure 4 of Ref. (*6*).

## Supercritical Isobutane Regeneration

Alkylation activity recovery from a completely deactivated USY zeolite catalyst was next examined in a single-pass reaction system employing SC

**Table II. Effect of Phase on TMP Regeneration Using $C_4$ and $C_5$ Alkanes at Near-critical Liquid and Supercritical Conditions**

| *Phase* | *n-butane* | *isobutane* | *n-pentane* | *isopentane* |
|---|---|---|---|---|
| | | *(% activity recovery)* | | |
| NC-L | 30 | 41 | 19 | 59 |
| SCF | 51 | 82 | 40 | 65 |

NOTE: Table based on Figure 5 of Ref. (*6*).

isobutane under variable regeneration conditions (*7*). Regeneration tests were performed to determine the effect of regeneration pressure, regeneration time, and regeneration temperature at constant pressure and at constant density, on catalyst activity recovery.

Regeneration temperature was varied while maintaining the reactor at the alkylation pressure of 111 bar. Temperatures tested (i.e., 60-190 °C) allowed for isobutane to be in the L, NC-L, and SCF phases as shown in Table III. Regeneration effectiveness increased with temperature up to an apparent maximum of 76–82% at 170-190 °C. These experiments, showed that merely washing the catalyst with isobutane in the liquid phase was ineffective at restoring catalyst activity. In contrast, regeneration conditions of either NC or SCF were necessary to obtain a significant recovery of catalyst activity. Increased temperatures under isobaric conditions increased both desorption rate of surface adsorbed species and solvent diffusivity, but decreased solvent density, and hence solute solubility. An effect of accelerated aging of coke may have also played a role in the decrease of regeneration effectiveness at the highest temperature applied.

When regeneration pressure was varied between 15-173 bar while maintaining the reactor temperature at 150 °C, a slightly lower TMP regeneration effectiveness was found at both the lowest and the highest regeneration pressures as shown in Table IV. This may be explained by the fact that at subcritical pressure, the solubility of carbonaceous compounds in the gaseous state is lower. As the pressure increases, the fluid density and therefore solute solubility increases, which increases regeneration effectiveness.

The slight decrease in regeneration effectiveness observed at the highest pressure studied may be the result of the combined effect of increased density with decreased diffusivity.

When regeneration time was varied between 30-180 min while maintaining the reactor at 150 °C and 111 bar, the majority of compounds that could be removed from the catalyst by SC isobutane were removed during the first 2 h of regeneration and 72% of the activity recovery that could be achieved in 2 h had been accomplished in the first 30 min of regeneration.

**Table III. TMP Regeneration Effectiveness for the Phase Tests**

| *Regeneration temperature (°C)* | *Regenerant phase* | *TMP regeneration effectiveness (%)* |
|---|---|---|
| 60 | L | 0.4 ± 0.6 |
| 90 | L | 2.8 ± 0.9 |
| 130 | NC-L | 41 ± 10.9 |
| 140 | SCF | 52 ± 1.6 |
| 150 | SCF | 61 ± 2.7 |
| 160 | SCF | 69 ±3.0 |
| 170 | SCF | 69 ±7.0 |
| 180 | SCF | 82 |
| 190 | SCF | 76 ±6.5 |

NOTE: Based on Table 2 and Fig. 6 of Ref (*7*).

**Table IV. TMP Regeneration Effectiveness as a Function of Pressure at 150 °C, Using 120 min Regeneration Time**

| *Regeneration pressure (bar)* | *15* | *42* | *70* | *111* | *139* | *173* |
|---|---|---|---|---|---|---|
| TMP reg. effectiveness (%) | 41 | 51 | 52 | 58 | 58 | 48 |

NOTE: Based on Figure 4 of Ref (*7*).

Another set of experiments tested the effect of temperature while maintaining a constant isobutane density of 0.41 g/cm$^3$. To obtain a constant density of the SC isobutene, pressure was adjusted at each temperature between 150 °C (111 bar) and 210 °C (220 bar). The results showed an apparent maximum in the TMP regeneration effectiveness between 190-210 °C, at about 78–83% recovered activity. Any increase in temperature increases all reaction rates and hence slow diffusing species may condense more rapidly and produce higher molecular weight hydrocarbons that could not to be extracted from the zeolite pores due to their larger size.

## Supercritical Fluid Regeneration of a Partially Deactivated USY Zeolite Catalyst

A summary of the research performed to explore two criteria to end the alkylation reaction and initiate SC isobutane regeneration on a partially deactivated USY zeolite catalyst is included in this section along with a summary of automated reaction/regeneration experiments aimed at determining the applicability of the process to refinery feed blends.

## Criteria for Initiation of Supercritical Isobutane Regeneration

The effectiveness of SC isobutane for single and multiple regenerations also was studied on a partially deactivated USY catalyst (***Error! Reference source not found.***). Catalyst samples were purposely deactivated to the point when either butene conversion declined to 95% or until a fixed (e.g., 3 hours) time-on-stream (TOS) was reached.

When the reaction was halted at a fixed TOS of 3.0 h, the integrated product yields were higher compared to those obtained from the fresh catalyst, the maximum product yields declined only slightly per run, and the overall product quality improved. Over 24 reaction cycles (see Table V), the TMP selectivity was relatively stable while dimethylhexane (DMH) selectivity decreased slightly. This provided an improvement in the TMP/DMH ratio from ca. 3.8 to ca. 5.0. This behavior, which was consistently observed, was mostly the result of higher product yields in the first hour of reaction likely due to the increased abundance of tert-butyl carbocations on the catalyst surface after regeneration.

By contrast, using the butene conversion bases for initiating regeneration the catalyst lost 48% of its initial TMP activity in just 9 reaction cycles. Cycle time decreased each run, declining from an initial value of 6.55 hours to a final time of 3.67 hours. Maximum instantaneous product yield also declined with each reaction cycle. These experiments demonstrated that high levels of catalyst activity recovery may be achieved if the level of deactivation prior to regeneration is limited and thus catalyst longevity can be increased by several orders of magnitude by using SC isobutane regeneration. Testing of repeated SC isobutane regeneration in a reaction/regeneration automated system on a commercially available refinery paraffin/olefin blend was studied next to analyze the potential of SC isobutane regeneration for industrial purposes. These studies are summarized below.

## Automated Reaction/Supercritical Isobutane Regeneration

Multiple supercritical isobutane regenerations of a partially deactivated USY solid acid catalyst also was tested utilizing a refinery alkylation feed blend (***Error! Reference source not found.***). The catalyst activity recovery was compared with the results of experiments that utilized a synthetic feed blend.

The refinery blend contained a variety of light olefins, and contaminants (e.g., butadiene, oxygenates, sulfur) which are known to produce significant catalyst deactivation. These experiments were performed in a continuous, automated reaction/regeneration system.

Thirty five 3-hour-reaction/3-hour-regeneration cycles were carried out over 210 h using the synthetic feed. High levels of butene conversion (i.e. above 92%) were maintained and product quality was relatively stable over the entire course of this experiment. The butene conversion declined an average of only 0.051% per cycle.

**Table V. Comparison of Product Quality Between Fresh Catalyst and a Catalyst Regenerated 24 Times After Halting the Reaction at 3 h TOS**

| | % of total alkylate | | | | | | | | |
|---|---|---|---|---|---|---|---|---|---|
| *Reaction Cycle No.* | $C_5$ | $C_6$ | $C_7$ | $C_8$ | *TMP* | *DMH* | *Other* $C_8$ | $C_{9+}$ | *TMP/DMH* |
| 1 | 7.6 | 5.1 | 7.1 | 76.9 | 59.8 | 15.6 | 1.5 | 3.4 | 3.84 |
| 24 | 6.8 | 4.7 | 7.1 | 78.3 | 61.7 | 12.4 | 4.2 | 3.2 | 4.97 |

NOTE: Table based on Table 3 of Ref. (*8*).

When the commercial feed was employed, high levels of alkene conversion were maintained for 78 h and 192 h using a 3-h-reaction/3-h-regeneration cycle and a 2-h-reaction/2-h-regeneration cycle, respectively. Employing a 3-h-reaction/3-h-regeneration cycle, alkene conversion remained at or above 92% for 13 reaction/regeneration cycles, corresponding to 78 h of semi-continuous operation. By utilizing a 2-h-reaction/2-h-regeneration cycle, alkene conversion did not drop below 92% until the end of 48 reaction/regeneration cycles, corresponding to 192 h of operation. The 2-h-reaction/2-h-regeneration cycle limited alkene conversion decline to 0.035% per run, and C8 and trimethylpentane (TMP) composition decline to 0.040% and 0.042% per run, respectively. Yields and productivities (defined as the product yield per gram of catalyst divided by the total TOS) are shown in Table VI for the control experiment (no regeneration) and for 2-h-reaction/2-h-regeneration cycle tests, using the commercial feed blend. Two criteria for ending the tests are shown: (1) Loss of 8 % of the alkene conversion and (2) Loss of 10% of the TMP activity since those were the points at which the regeneration tests were terminated.

When 8% loss of alkene conversion was used as the criterion for comparison, the regenerated swing reactors combined for 23 fold increases in productive time on stream, for the 2h/2h tests, versus the non-regenerated control case. The cumulative $C_{5+}$ and TMP yields were increased by 24 fold and 22 fold, respectively. The $C_{5+}$ and TMP productivities were essentially unchanged versus the control, with slight increases for $C_{5+}$ and slight decreases for TMPs.

When 10% loss of TMP activity was used as the criterion for comparison, the regenerated swing reactors combined for 31 fold increases in productive time on stream for the 2h/2h tests, versus the non-regenerated control case. The cumulative $C_{5+}$ and TMP yields were increased by 70 fold and 59 fold, respectively. Using this criterion, both the $C_{5+}$ and TMP productivities increased by a factor of about 2 fold versus the control productivity. These results were highly encouraging since they demonstrated the net effect of SC isobutane regeneration in extending the active catalyst lifetime, thereby allowing higher yields from a unit of catalyst when refinery paraffin/olefins blend were utilized.

**Table VI. Comparison of Cumulative Yields and Productivities for Cycle Testing Using a Commercial Feed Blend.**

| | *No Regeneration* | *2-h-React/2-h-Regen Cycle* |
|---|---|---|
| Criterion: 8% loss of alkene conversion | | |
| TOS (h) | 8.34 | 192.3 |
| Cumulative $C_{5+}$ Yield (g/g cat) | 1.303 | 30.72 |
| Cumulative TMP Yield (g/g cat) | 0.601 | 13.15 |
| $C_{5+}$ Productivity (g/g cat/h) | 0.156 | 0.160 |
| TMP Productivity (g/g cat/h) | 0.0720 | 0.0684 |
| Criterion: 10% loss of TMP activity | | |
| TOS (h) | 3.33 | 104.2 |
| Cumulative $C_{5+}$ Yield (g/g cat) | 0.251 | 17.52 |
| Cumulative TMP Yield (g/g cat) | 0.130 | 7.67 |
| $C_{5+}$ Productivity (g/g cat/h) | 0.0754 | 0.168 |
| TMP Productivity (g/g cat/h) | 0.0392 | 0.0736 |

NOTE: Based on Table 4 of Ref. (*9*).

## Characterization of Spent Zeolite Catalysts

This section summarizes the chemistry of the SC isobutane regeneration process. To understand the nature of the hydrocarbons that remain adsorbed on the surface of the USY zeolite catalyst both before and after SC isobutane regeneration, a series of ex-situ temperature-programmed oxidation (TPO), diffuse reflectance infrared Fourier transform spectroscopy (DRIFTS), and ultraviolet-visible (UV–vis) analyses was performed on samples submitted to different TOS (*10*) under isobutane/butene reaction conditions.

These analyses demonstrated that during alkylation, the adsorption of hydrocarbons, which were mostly highly branched paraffins, decreased catalyst micropore volume, and that fully deactivated USY samples did not present any available microporosity. SC isobutane was effective in recovering catalyst microporosity even from the completely deactivated USY samples. However, full removal of hydrocarbon species was only obtained from samples that were not completely deactivated. Samples that were initially utilized for long TOS contained highly condensed carbonaceous species that instead of being extracted by the SC isobutane, dehydrogenated to produce species even more condensed. This further condensation during SC isobutane regeneration was assigned to the higher temperature (i.e., 180 °C) used for regeneration with respect to the milder conditions (i.e., 60 °C) applied during alkylation.

To determine the effect of zeolite pore structure on coke removal by SCF regeneration a series of wide pore zeolites, which included acidic Ys, betas, and mordenite, were fully deactivated under a flowing isobutane/butene mixture and then regenerated under flowing SC isobutane for 60 min (*11*). The spent catalysts were examined ex-situ by TPO, DRIFTS, and UV-Vis spectroscopy both before and after regeneration. These analyses demonstrated that although most adsorbed hydrocarbons were removed in some catalysts none of the catalysts were completely free of hydrocarbon deposits after SC isobutane regeneration.

Zeolite pore structure and acidity were shown to play important roles in hydrocarbon transformation and removal by SC isobutanc regeneration. The lack of complete regeneration of fully deactivated samples was assigned to the formation of condensed hydrocarbon species. This effect was favored by the presence of cavities in the zeolite pore structure, lower zeolite acidities, and higher temperatures during SC isobutane regeneration with respect to the temperature during isobutane/butene reaction.

Zeolites possessing high acidities and three-dimensional pore structures that lack large expansions or cavities are likely the most adequate for repeated SC isobutane regeneration. These type of zeolites are expected to favor hydride transfer rates during SCF regeneration and impose steric hindrance on the formation of condensed hydrocarbon species. This was confirmed during cyclic reaction/SCF regeneration experiments on fully deactivated USY and beta zeolites. Although neither catalyst recovered its initial performance after SC isobutane treatment, the decrease of activity between cycles was slower in the beta sample likely due to hindered formation of condensed compounds.

## Conclusions

Supercritical isobutane can provide complete activity recovery from partially deactivated USY catalysts. It can also be applied multiple times to maintain high levels of catalyst activity when the alkylation reaction is performed utilizing commercially available refinery blends. Pressure, temperature, and regeneration time played important roles in the SC isobutane regeneration process because of their effect on solute solubility, diffusivity, surface desorption, hydride transfer rates, and coke aging. As a consequence, regeneration effectiveness may be maximized by manipulating those variables.

Zeolite pore structure and acidity also play important roles in hydrocarbon transformation and removal by SC isobutane regeneration. The lack of complete regeneration of fully deactivated samples was attributed to the formation of condensed hydrocarbon species. This effect was favored by the presence of cavities in the zeolite pore structure, lower zeolite acidities, and higher temperatures during SC isobutane regeneration with respect to the temperature during isobutane/butene reaction. Zeolites possessing high acidities and three-dimensional pore structures that lack large expansions or cavities are likely the most adequate for repeated SC isobutane regeneration. These types of zeolites are expected to favor hydride transfer rates during SCF regeneration and impose steric hindrance on the formation of condensed hydrocarbon species.

In summary, the research work performed at the INL demonstrated that the introduction of supercritical cosolvents during the alkylation reaction did not result in improved or sustained catalytic performance. However, supercritical fluids in general and isobutane in particular were shown to be promising regenerants of some solid acid zeolite catalysts that may be utilized in isobutane/butene alkylation reaction.

## Acknowledgement

This work was supported by the US Department of Energy, Office of Fossil Energy, National Energy Technology Laboratory and INL Laboratory Directed Research and Development Program, under contract number DE-AC07-05ID14517. In-kind financial contributions to this work by Marathon Ashland Petroleum, LLC, through a Cooperative Research and Development Agreement is acknowledged.

## References

1. Albright, L. F. *Oil Gas J.* **1990**, *88*, 79–&.
2. Corma, A.; Martinez, A. *Catal. Rev.-Sci. Eng.* **1993**, *35*, 483–570.
3. Weitkamp, J.; Traa, Y. *Catal. Today* **1999**, *49*, 193–199.
4. Baiker A. *Chem. Rev.* **1999**, *99*, 453–473.
5. Ginosar, D. M.; Thompson, D. N.; Coates, K.; Zalewski, D. J. *Ind. Eng. Chem. Res.* **2002**, *41*, 2864–2873.
6. Ginosar, D. M.; Thompson, D. N.; Burch, K. *Appl. Catal. A* **2004**, *262*, 223–231.
7. Thompson, D. N.; Ginosar, D. M.; Burch, K. C. *Appl. Catal. A* **2005**, *279*, 109–116.
8. Thompson, D. N.; Ginosar, D. M.; Burch, K. C.; Zalewski, D. J. *Ind. Eng. Chem. Res.* **2005**, *44*, 4534–4542.
9. Ginosar, D. M.; Thompson, D. N.; Burch, K. C. *Ind. Eng. Chem. Res.* **2006**, *45*, 567–577.
10. Petkovic, L. M.; Ginosar, D. M. *Appl. Catal. A* **2004**, *275*, 235–245.
11. Petkovic, L. M.; Ginosar, D. M.; Burch, K. C. *J. Catal.* **2005**, *234*, 328–339.

# Chapter 14

# ExSact: Novel Solid-Acid Catalyzed Iso-Paraffin Alkylation Process

**Mitrajit Mukherjee, Sankaran Sundaresan, Richard Porcelli, and James Nehlsen**

**Exelus Inc., 99 Dorsa Avenue, Livingston, NJ 07039**

The first commercially viable solid-acid catalyst for isoparaffin alkylation is developed by simultaneously applying basic engineering concepts at many scales during process development. The pore structure, the acid site density, and the active site distribution are optimized to achieve the desired isomeric products, while resisting coking. Concurrent reactor design further enhances product quality, while reducing capital cost. The result is a "green" process that eliminates the hazards of liquid-acid catalysts, while also reducing cost.

The phase-out of methyl tertiary-butyl ether (MTBE) has important consequences on the US reformulated gasoline (RFG) market both in terms of octane value and volume loss. MTBE has an octane rating of 110 and it commands approximately 10% of the RFG and California Air Resources Board (CARB) gasoline pool. Although MTBE will likely continue to be used in the rest of the world, eliminating it from domestic gasoline will require compensation in the form of alternative high-octane additives. The shortfall in gasoline volume and octane rating can be met by substituting alkylate for MTBE.

Alkylate is an ideal clean fuel component because it has a high octane rating, low vapor pressure, low toxicity, and costs less than MTBE. Table 1 compares some relevant properties of alkylate and MTBE. Alkylate has been blended into gasoline for decades to improve octane. EIA (Energy Information

Table I. Properties of Alkylate compared with MTBE.

| Selected High – Octane Components | RVP (psia) | Octane Number (R+M)/2 | Relative Cost |
|---|---|---|---|
| MTBE | 8.0 | 110 | 1.6 |
| Alkylate | 2.6 | 94 | 1.0 |

Administration) reported *(1)* that the volume of total gasoline production can be restored to MTBE-free volume levels with alkylate. Refiners, however, are reluctant to replace MTBE units with costly liquid acid-catalyzed alkylation units, which pose severe environmental hazards and security risks.

Alkylate is currently produced using either sulfuric or hydrofluoric acid as the catalyst. HF is particularly dangerous due to its propensity to form stable aerosols in an accidental release. Sulfuric is less dangerous, but its use still requires brick-lined vessels and consumption is high. Sulfuric acid units also require costly refrigeration to maintain the proper reaction temperature.

Solid acid-catalyzed alkylation eliminates the hazards and costs associated with using and regenerating corrosive liquid acids. Unfortunately, most solid acid technologies are significantly more expensive than liquid acid systems and require complex reactor systems to compensate for short catalyst life. As a result, grassroots systems are not typically competitive with liquid acid technology, and retrofitting is often not possible due to the exotic process designs. Of the four solid acid alkylation technologies proposed by major licensors, none has yet been accepted by the industry.

MTBE producers wanting to switch production to alkylate have been restricted primarily to isooctene (isobutene dimerization) technology. Isooctene has a much lower motor octane number (MON) value than paraffinic alkylate and requires an expensive hydrogenation reactor to create high quality product. Solid acid-catalyzed alkylation technology compatible with existing packed bed reactors resolves all of these issues. Both retrofitting and grassroots plants become economical. Process operation is simple, and product quality from solid acids can equal or exceed that of liquid acid processes. Alkylation also provides more than twice the volume yield of dimerization, restoring capacity lost by eliminating methanol as a feed stream in MTBE facilities.

## Alkylation Reactions

The basic reactions occurring during isoparaffin alkylation are shown in figure 1. Alkylation of isobutane with light ($C_3$ –$C_5$) olefins in the presence of a strong acid catalyst involves a series of consecutive and simultaneous reactions occurring through carbocation intermediates *(2)*. The protonation of an olefin,

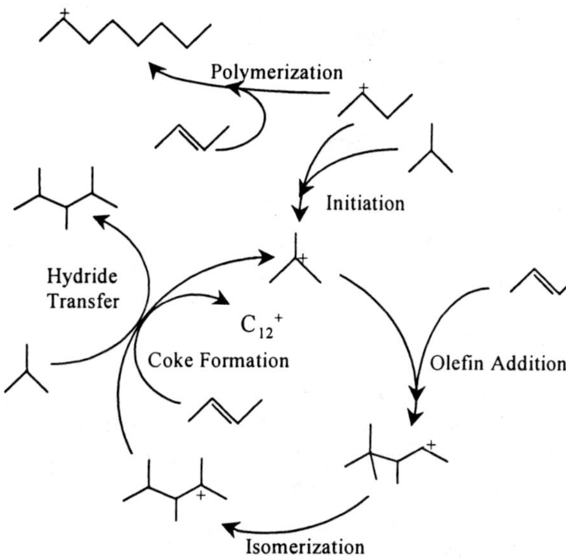


*Figure 1. Isoparaffin alkylation reactions in the presence of an acid catalyst.*

followed by hydride abstraction from isobutane, leads to a *t*-butyl cation and initiates the reaction. The *t*-butyl cation then combines with an olefin (e.g. $C_4$) to give the corresponding $C_8^+$ carbocation. These $C_8^+$ carbocations may isomerize via hydride transfer and methyl shifts to form more stable cations. They also undergo hydride transfer with isobutane to regenerate the *t*-butyl cation and perpetuate the chain sequence.

Unfortunately, these are not the only reactions occurring during alkylation. There are a number of secondary reactions that, in general, tend to reduce the quality of alkylate. The $C_8^+$ can continue to react with olefins to form larger cations. The successive addition of olefins to $C_8^+$ carbocations and olefin polymerization are the primary routes to formation of high molecular weight hydrocarbon residue or "coke." Liquid-acid catalysts generate heavy byproducts in the form of "acid-soluble oil." This byproduct must be treated to remove the hydrocarbon content and regenerate the acid, leading to higher catalyst costs, particularly for sulfuric acid systems. Solid-acid catalysts offer a safe and environmentally benign alternative to liquid acids. However, their fine pore structure is prone to clogging by these large coke molecules, and the acid sites can be deactivated by strongly adsorbed species. Most solid acid catalysts therefore deactivate quickly unless the olefin concentration in the reactor is kept to extremely low levels *(3)*. Most solid acid catalysts available today last only an hour or two at typical reaction conditions.

## Criteria for a Viable Process

### *Catalyst*

Successful design of an isoparaffin alkylation catalyst must overcome the following two challenges:

- The catalyst must facilitate the hydride transfer reaction both during the initiation reaction and, more importantly, the alkylation step.
- The formation of coke precursors, including olefin oligomers, must be suppressed to extend catalyst lifetime.

Most strong Brønsted acids effectively facilitate the hydride transfer reaction required to make alkylate *(4)*, but avoiding the formation of high molecular weight coke precursors has proven more difficult to achieve. This second hurdle is particularly important for the deactivation of solid acid catalysts and has proven to be a stumbling block for the technology.

Any catalyst chosen should also meet certain criteria from a practical standpoint. The catalyst should be physically and chemically stable to ensure product purity and catalyst lifetime. Supported liquid acid species are undesirable due to leaching, and many Lewis acids, such as chlorided alumina, require a co-feed of halogen to compensate for chemical degradation caused by water. A true solid acid is the only catalyst that is acceptably stable.

### *Reactor*

The key challenge in isoparaffin alkylation is to create a catalyst/reactor system that minimizes the rate of deactivation of the catalyst. To maintain economic viability, the system should not be overly complex or expensive.

Alkylation is a fast reaction over an appropriate catalyst. In a simple packed bed reactor, the reaction zone is very narrow and catalyst deactivation is rapid. A reaction front is created where most of the olefin present in the feed reacts in a small active zone. Behind the front, the catalyst is deactivated. Ahead of the front, the olefin is depleted and only cracking reactions can occur.

The front moves through the bed as the catalyst deactivates. Breakthrough of the olefin occurs when the front reaches the end of the bed. This effect is diagrammed in figure 2. The same effect occurs in aromatic alkylation, though the deactivation is much slower. A packed bed will rapidly deactivate since the olefin concentration at the front is very high, equal to the isoparaffin to olefin ratio (I/O) of the feed unless a recycle is used. With conventional catalysts in a packed bed, the front will move through the bed rapidly, often in less than one hour, far too fast for commercial application.

The deactivation of solid acid catalysts occurs through the formation of large hydrocarbons that cannot effectively desorb from the active site and diffuse out of the catalyst. These species are formed either by the addition of olefins to $C_8$ and heavier molecules or through olefin oligomerization. The kinetic rate constant for olefin oligomerization in most acids is at least two orders of magnitude higher than that for the hydride transfer reaction required to form alkylate. The concentration of olefin at the catalyst surface should, therefore, be at least two orders of magnitude lower than the isoparaffin concentration to enable the alkylation reaction to proceed faster than oligomerization. Reducing the concentration of olefin in the reactor is well known to reduce the formation of coke and extend catalyst life. *(5)* Figure 3 shows that an I/O ratio in the reactor of about 400 is needed to prevent rapid deactivation. However, feed I/O ratios significantly lower (typically 6-15) are required for economical separation after the reactor. The reactor design must, therefore, create high I/O ratios in the reactor despite low feed ratios.

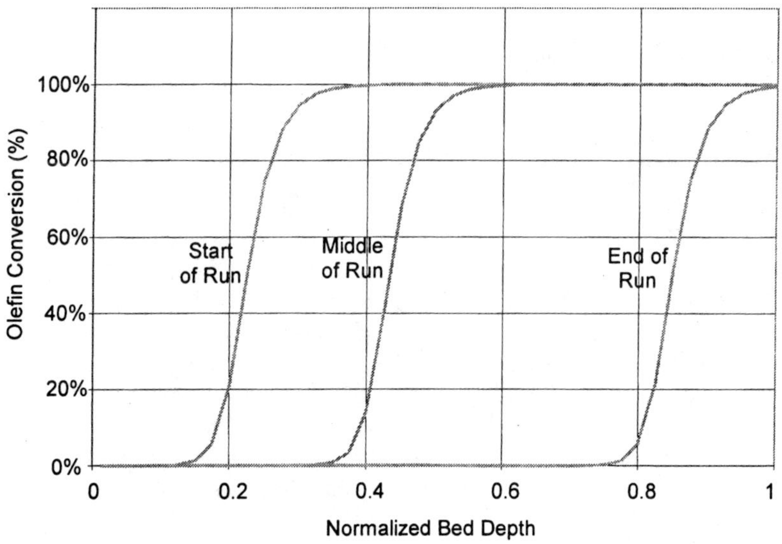


*Figure 2. Hypothetical reaction front moving through a packed bed.*

In response to this limitation, several processes have been developed that suppress catalyst deactivation through the use of slurry reactors, catalytic distillation, and an FCC-style fluidized bed with constant catalyst regeneration *(6-8)*. These reactors achieve high I/O ratios and lower catalyst deactivation rates, but at a cost. Fluidized beds, for example, require specialized, attrition-resistant catalyst and extensive filtration equipment.

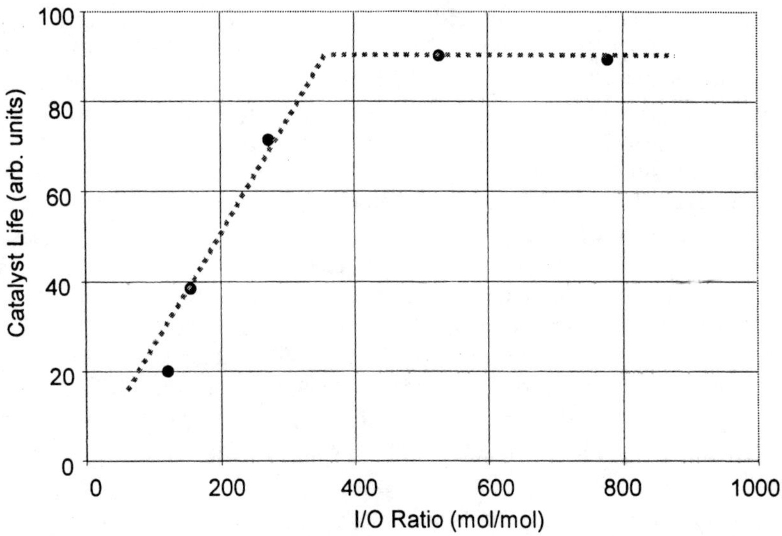


*Figure 3. Effect of isoparaffin to olefin (I/O) ratio on catalyst lifetime.*

These reactor types also inevitably suffer from side reactions. Specifically, when alkylate concentrations are high, as is the case in any well-mixed reactor, alkylation of the $C_8$ products causes product degradation and catalyst deactivation. The rate of this reaction is comparable to the rate of olefin addition to isobutane, since both are isoparaffins. Long residence times result in cracking of alkylate into undesirable light components, further reducing product yield.

These side reactions suggest two conditions to be met by the reactor system. A high I/O ratio promotes longer catalyst life by reducing the formation of heavy byproducts, and a high isobutane to alkylate ratio promotes high product octane by reducing cracking reactions. A well-mixed reactor achieves the first of these conditions, but not the second. A packed bed achieves the second, but not the first. Achieving both conditions requires a different approach.

## Engineering a Catalyst

Using grants from the US Department of Energy and the National Science Foundation, Exelus started developing a viable solid-acid catalyst alternative for isoparaffin alkylation. In order to produce a commercially viable process, the goal was to develop a catalyst that was useable in a simple fixed-bed reactor process using a benign solid-acid catalyst. Our desire to use a fixed-bed reactor was influenced by two issues: a) to simplify process scale-up and b) to keep both capital costs and operating expenses as low as possible. In order to achieve this level of performance, every aspect of the catalyst must be engineered to meet the reaction criteria. This, however, is easier said than done for alkylation reactions.

Isoparaffin alkylation reactions are very fast and they suffer from severe pore diffusion limitations. As a result, when catalyst particle size is increased from 100 μm (for slurry reactors) to 1.6 mm for fixed-bed reactors, the catalyst activity reduces by 10-fold according to basic mass transfer models using experimental values of the intrinsic rate constant, as shown in figure 4. To match the catalyst productivity of a slurry reactor, one would need to build a fixed-bed reactor with ten times the volume – not practical for a commercial scale system. In addition to using a fixed-bed reactor, we wanted to ensure that the solid-acid catalyst was both robust as well as benign (i.e. environmentally friendly).

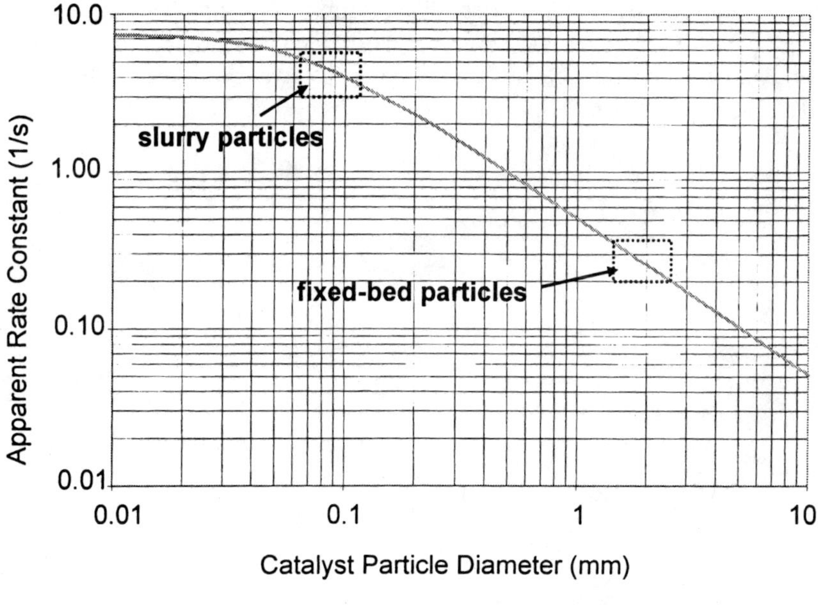


*Figure 4. Reduction in kinetic rate constant due to mass transfer limitations.*

In order to create a proper catalyst, it is necessary to first understand the dynamics of the alkylation reaction by systematically studying the effect of both physical and chemical aspects of the solid-acid catalyst system design on product distribution and catalyst deactivation characteristics. Several different catalyst formulations encompassing a wide range of acid functions, density and distribution of acid sites, and acid strengths were evaluated as part of this study.

Through this systematic study, Exelus was able to identify an optimal "window" of design parameter values that were then used to develop the catalyst. By judicious manipulation of the active material composition, researchers at Exelus developed a unique solid-acid catalytic system that has roughly 400% more active sites than a typical solid-acid catalyst. The catalyst activity was found to be higher than a typical liquid acid catalyst, which means that smaller amounts of catalyst are required, allowing one to design alkylation reactors with significantly lower volumes.

By integrating optimized acid sites with superior mass transport characteristics and a pore architecture that reduces pore-mouth plugging, a catalyst with enhanced performance can be created. Figure 5 demonstrates that both the catalyst selectivity and lifetime are significantly improved. As shown in figure 6, which compares the performance of Exelus solid-acid catalyst with other commercially available systems, the new catalyst system is easily able to achieve a step-change in performance over other solid-acid catalysts.

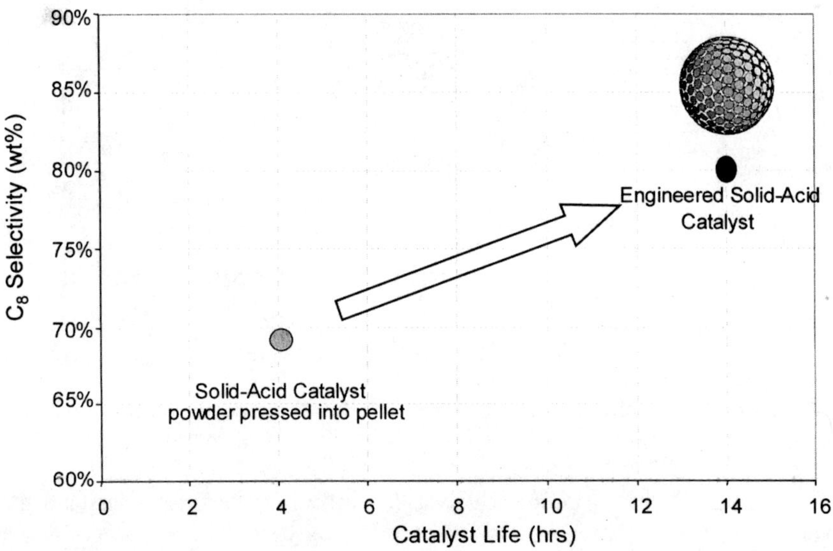


*Figure 5. Improvement in catalyst performance as a result of engineered design.*

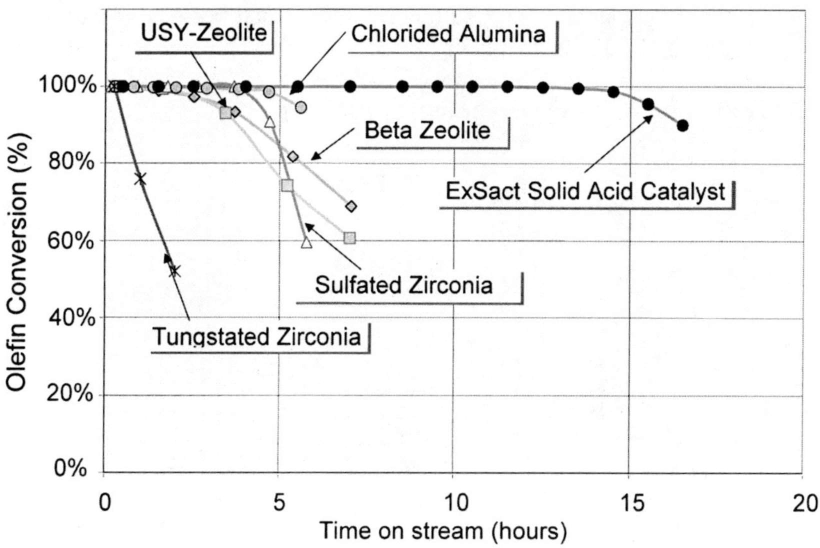


*Figure 6. Relative performance of an engineered catalyst to other solid acid catalysts.*

## Novel Fixed-Bed Reactor Design

Through an extensive study of the solid-acid alkylation reactions, Exelus has developed a robust kinetic model that is able to predict product octane as well as catalyst lifetimes for a large range of reaction conditions. Kinetic analyses *(9)* revealed that the rate constant for the undesired secondary (oligomerization) reaction is 2 orders of magnitude higher than the rate constant for the primary alkylation reaction. To achieve a substantial reduction in the rate of oligomerization, the olefin concentration (O) must therefore be kept 2 orders of magnitude lower than the isoparaffin concentration (I). Kinetic analyses also revealed that the rate constant for the product degradation reaction is similar in magnitude to the rate constant for the primary alkylation reaction.

Using these data, a kinetic model capable of predicting the product yield and quality could be developed. By applying the kinetic model to novel reactor flow patterns, Exelus has been able to develop a unique fixed-bed alkylation reactor that enhances the primary reaction rate by optimizing both macro-mixing (residence time distribution) and micro-mixing (fluid dispersion) characteristics. The macro-mixing profile is optimized to reduce product degradation by minimizing the over-alkylation of iso-octane molecules to heavier (low octane) hydrocarbons. The desired micro-mixing characteristics produce low olefin concentrations at the catalyst surface by dispersing small pockets of olefin near the injection site. This effect is achieved by using proprietary turbulence enhancing devices called "Dispersion Accelerators." The innovations in the Exelus process are summarized schematically in figure 7.

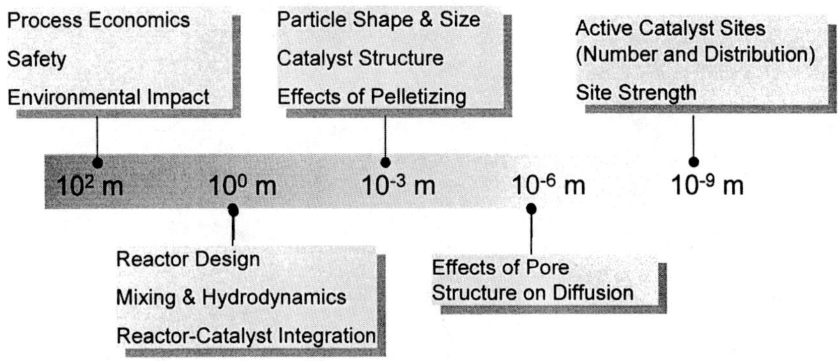


*Figure 7. The design of the alkylation process must occur on multiple length scales simultaneously.*

The performance of the fixed-bed reactor was verified in Exelus laboratory over a period of 2 months through tests under commercially relevant process conditions. Product was sampled over the entire length of the reactor to quantify the rate of alkylate degradation. Figure 8 shows the results of this test.

The kinetic model adequately describes the product composition for most of the reactor length. In effect, the yield of $C_8$ alkylate is 20 wt% higher than that possible in a conventional recycle reactor, resulting in a net 1.5 octane number (RON) increase.

## Comparison with Conventional Liquid Acid Processes

A comparison of the Exelus technology with competing liquid acid processes is presented below. We have used published data and pro-rated them (the sulfuric acid unit *(7)* was pro-rated from a 7,500 barrel per day (bpd) plant cost of $43.5 million while the HF plant *(6)* was pro-rated from a 8,800 bpd plant cost of $42.5 million) to reflect a plant capacity of 10,000 bpd.

Capital expenses for the new solid acid process are lower than liquid-acid technologies due to the following reasons:

- Simpler reactor design using inexpensive materials of construction
- Elimination of refrigeration equipment
- No need for acid handling vessels

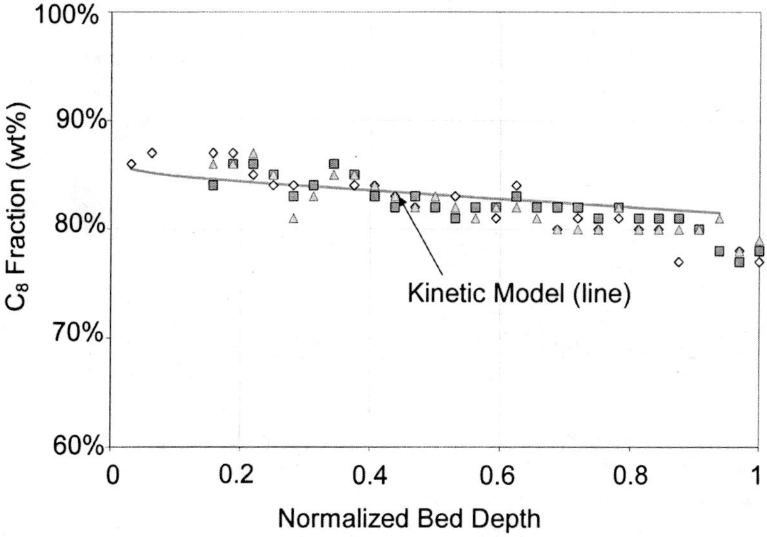


*Figure 8. Fit of kinetic model to optimized reactor product quality.*

**Table II. Comparison of Capital and Operating Costs of Liquid and Solid Alkylation Technologies**

| Parameter | ExSact Process | Hydrofluoric Acid Alkylation | Sulfuric Acid Alkylation |
|---|---|---|---|
| Plant Capacity (bspd) | 10,000 | 10,000 | 10,000 |
| **Capital Expenses (million)** | **$ 26** | **$ 46** | **$ 55** |
| **Raw Material Consumption** : | | | |
| Iso-butane Consumption | 1.21 v/v olefin | 1.2 v/v olefin | 1.17 v/v olefin |
| Alkylate Yield | 1.83 v/v olefin | 1.8 v/v olefin | 1.78 v/v olefin |
| ***Utility/Chemicals Consumption*** : | | | |
| Steam ($ 4/klb) | 116,000 lb/hr | 65,700 lb/hr | 130,700 lb/hr |
| Power ($ 0.06/kWH) | 600 kW | 903 kW | 4040 kW |
| Fuel ($ 4/MM Btu) | 1.5 MM Btu/hr | 47.7 MM Btu/hr | |
| Cooling Water ($ 0.05/gal) | 12,740 gpm | 15,750 gpm | 15,600 gpm |
| Hydrogen ($ 0.55/lb) | 70 lb/hr | | |
| HF ($ 0.65/lb) or $H_2SO_4$ ($ 0.05/lb) | N/A | 47 lb/hr | 7920 lb/hr |
| **Utility Costs ($/bbl)** | **$ 1.40** | **$ 1.41** | **$ 2.90** |

As a result, capital expenses are about $26 million for a 10,000 bpd plant – a reduction of 45-50% compared with hydrofluoric and sulfuric acid alkylation processes. Utility costs are about the same as an HF unit but significantly lower (about 50%) than a sulfuric acid plant due to much lower power consumption and an inexpensive regeneration scheme using hydrogen. Finally, alkylate yields in the solid acid technology are higher due to elimination of acid-soluble oils that are formed during liquid-acid catalyzed alkylation processes, resulting in lower feedstock consumption.

## Conclusions

A solid acid catalyst for isoparaffin alkylation has been engineered to produce high-octane alkylate without the use of toxic liquid acids. The strength and distribution of its acid sites reduce olefin dimerization and paraffin cracking, while enhancing isobutane alkylation. Furthermore, the controlled pore structure of the catalyst reduces deactivation by coke formation and pore mouth plugging. Both the active catalyst material and the overall structure and design of the catalyst are engineered to yield dramatic improvements in product quality and catalyst lifetime. The isoparaffin alkylation process using this catalyst is fundamentally safer and cleaner than conventional liquid-acid catalyzed processes, eliminating the use and generation of toxic chemicals.

A novel fixed-bed alkylation reactor co-designed with the catalyst further enhances product quality, reduces deactivation, and minimizes capital and operating costs. This new reactor uses innovative fluid dynamics and an unusual reactor residence time distribution designed to maximize alkylate production without sacrificing the product octane value. The new solid-acid catalyst process promises significantly improved yields and selectivity while minimizing waste by-products and disposal problems associated with spent catalysts and regeneration of large quantities of liquid acids. Given the current push for eco-friendly processes to produce ultra-clean fuels, this type of technology is poised to fill a large void in the gasoline market for decades to come.

## Acknowledgements

This work was funded by grants from the U.S. Department of Energy and the National Science Foundation.

## References

1. Energy Information Administration, *Supply Impacts of an MTBE Ban.* September 2002.

2. Feller, A., et al., Journal of Catalysis, 2003. **216**: p. 313-323.
3. Corma, A. and A. Martinez, Catalysis Review Science and Engineering, 1993. **35**: p. 483.
4. Roper, M., et al., *Acylation and Alkylation*, in *Ullman's Encyclopedia of Industrial Chemistry*. 2003, Wiley-VCH: Weinheim. p. 337-382.
5. Sahebdelfar, S., et al., *Deactivation behavior of the catalyst in solid acid catalyzed alkylation: effect of pore mouth plugging.* Chemical Engineering Science, 2002. **57**: p. 3611-3620.
6. UOP LLC., *HF Alkylation.* 2001.
7. Ackerman, S., G.K. Chitnis, and D.S. McCaffrey. *ExxonMobil Sulfuric Acid Alkylation Process.* in *222nd ACS National Meeting*. 2001. Chicago.
8. Ondrey, G., A new alkylation process to be less expensive, problematic in *Chemical Engineering*, 7, 2003, pp.17.
9. Simpson, M.F., J. Wei, and S. Sundaresan, *Kinetic Analysis of Isobutane/Butene Alkylation over Ultrastable H-Y Zeolite.* Industrial Engineering and Chemistry Research, 1996. **35**: p. 3861-3873.

# Utilization

# Chapter 15

# Screening of Potential O-Ring Swelling Additives for Ultraclean Transportation Fuels

**John P. Baltrus[1], Dirk D. Link[1], Paul H. Zandhuis[2], Robert J. Gormley[1], and Richard R. Anderson[1]**

[1]National Energy Technology Laboratory, U.S. Department of Energy, P.O. Box 10940, Pittsburgh, PA 15236–0940
[2]National Energy Technology Laboratory Site Support Contractor, Parsons RDS, P.O. Box 618, South Park, PA 15129

Several classes of organic compounds and mixtures of organic compounds were evaluated as potential additives to Fischer-Tropsch fuels to promote swelling of nitrile rubber o-rings that come in contact with the fuels. Computational modeling studies were also carried out to predict which compounds might be best at promoting o-ring swelling. The combined experimental-theoretical approach showed that steric factors strongly influence the interactions between additives and the nitrile sites in the rubber that result in swelling. Select compounds incorporating both oxygenate and aromatic functionalities appear to be the best candidates for additives because of a "dual" interaction between complementary functionalities on these compounds and the nitrile rubber.

The primary components of Fischer-Tropsch (F-T) fuels are typically linear and branched alkanes (*1*), in addition to smaller amounts of other compounds such as alkenes depending on the process that was used to produce the fuels. These compounds have a very weak ability to swell nitrile rubber o-rings. Such o-rings are the predominant type of o-rings found in legacy ground and air vehicles used in military operations as well as in civilian transportation vehicles. Under normal conditions, nitrile rubber o-rings will swell when they come in contact with a petroleum-derived fuel and this swelling is taken into account in designing systems that require o-rings to form a tight seal to prevent fuel leakage.

Because of the increasing cost of petroleum-derived fuels and the desire to rely less on non-domestic sources of those fuels, there has been a renewed interest in incorporating F-T fuels into the domestic transportation fuel supply inventory. Before that can take place, issues such as the potential shrinkage of nitrile rubber o-rings when they contact fuels of increasing F-T character must be addressed. Otherwise there is the potential for disastrous consequences due to o-ring shrinkage and subsequent fuel leakage.

Two of the most important properties of F-T fuels that need to be addressed before their incorporation into the transportation fuel supply stream are lubricity and seal swelling capability (*2-4*). A near-term solution to the seal swelling and lubricity deficiency of F-T fuels is to blend the F-T fuel with conventional petroleum-derived feedstocks (*4*). Unfortunately that solution diminishes many of the benefits derived from using a pure ultra-clean transportation fuel. For the same reason, addition of sulfur compounds, which would induce desirable lubricity properties, is not an option. Therefore other classes of compounds, specifically aromatics and oxygenates, are being investigated as additives. This study seeks to determine which classes of compounds most efficiently promote nitrile rubber o-ring swelling. Of particular interest is the determination of what specific compounds would be the best swelling additives.

## Experimental

### Swelling Studies

The swelling properties of the fuels were measured using ASTM Methods D-1414 and D-471. The o-rings used in the studies were buna-N (nitrile) N0602-70, obtained from Parker Seals, and having an outer diameter and thickness of approximately 1.9 mm and 0.3 mm, respectively. Measurements were simultaneously conducted on three o-rings, as prescribed by the ASTM methods. The volume of fuel used was typically 40 mL. The concentrations of additives are reported as volume percent, unless indicated otherwise. The immersion time for the o-rings was 48 hours.

### Compositional Analysis

Analysis was performed with an Agilent gas chromatography/mass spectrometry (GC/MS) system (Agilent 6890 GC coupled to an Agilent 5973 Mass Selective Detector). Depending on the information desired, either a non-polar column (Petrocol-DH 50.2, Supelco) or a polar column (Stabilwax-DA, Restek) was used. The polar column was used primarily for the analysis of extracts to help isolate polar components from the non-polar bulk matrix. Peak identification was performed using Agilent's ChemStation processing software against a National Institute of Standards and Technology (NIST) library of compounds, with additional confirmations obtained using the Automated Mass Spectral Deconvolution and Identification System (AMDIS) software package.

### Computational Studies

Computational modeling of the interactions between specific organic molecules and the isobutyronitrile functionality was carried out using a scheme adapted from previous work conducted at the National Energy Technology Laboratory (NETL) (*6*). In a similar but slightly modified approach, *ab initio* calculations using the Gaussian 03 package (*7*) were performed to determine optimized structures at the second order Møller-Plesset (MP2) level of theory using the cc-pVDZ basis set for organic molecules and the isobutyronitrile monomer, model polymer. Fully optimized combined structures for each of these molecules with the isobutyronitrile monomer were calculated with the same cc-pVDZ basis set and at the MP2 level of theory. Finally, single point energy counterpoise calculations were made to determine the Basis Set Superposition Error (BSSE) corrected interaction energies for these molecules with the model polymer.

## Results and Discussion

Base o-ring swelling measurements were conducted on petroleum-derived and F-T diesel and jet fuels to determine the extent of swelling problems that might be encountered in moving from a petroleum-derived fuel to a synthetic fuel. The results are reported in Table I. As expected, the petroleum-derived fuels promoted excellent nitrile o-ring swelling while the F-T fuels yielded poor swelling results. Treatment in F-T diesel fuel actually resulted in shrinkage of the o-rings.

**Table I. Swelling Results for Nitrile Rubber O-Rings in Various Fuels**

| *Fuel* | *Volume Change (%)* |
|---|---|
| Commercial Diesel | 16.2 ± 1.0 |
| JP-5 | 14.0 ± 0.6 |
| F-T Diesel | - 0.7 ± 0.3 |
| F-T Jet (S-5) | 0.7 ± 0.2 |
| S-5 + 10% EI-037 | 8.5 ± 0.4 |
| S-5 + 15% EI-037 | 13.6 ± 0.2 |
| S-5 + 25% EI-037 | 25.4 ± 0.8 |
| S-5 + 15% Aromatic 150 | 9.6 ± 1.3 |

NOTE: Vales are reported with a 95% confidence interval.

A previous study attempted to determine the appropriate levels for spiking a synthetic jet fuel (S-5) with an aromatic-containing petroleum-derived jet fuel (JP-5) or an Aromatic solvent mixture, Aromatic-150 (ExxonMobil), to induce an acceptable level of swelling for nitrile rubber o-rings (*8*). It was determined that JP-5/S-5 blends were more proficient than Aromatic-150/S-5 blends at swelling nitrile rubber o-rings at a given level of aromatic content for the blend. Specifically, a 30 % blend of Aromatic-150 in S-5 was required to induce the same degree of swelling that was induced by a JP-5 fuel containing 20% aromatics (*8*). The results reported in Table I show the same behavior, but less dramatically. Also, a highly-aromatic coal-derived jet fuel precursor, EI-037 (Energy Research Institute, Pennsylvania State University), was found to induce significant swelling of nitrile rubber o-rings. The data in Table I show approximately the same level of swelling for the JP-5 fuel and an S-5/EI-037 blend where the EI-037 concentration is only 15%. These results indicate that while aromatic species are able to induce swelling, some species (or combination of species) present in the petroleum-derived and coal-derived fuels are more proficient at inducing swelling than the aromatic species in the Aromatic-150 solvent.

A comparison of Aromatic-150 and JP-5 was made using GC/MS with a non-polar analytical column. The results showed that the region of overlap between Aromatic-150 and JP-5 ended at approximately the point where dodecane eluted. Therefore, it was inferred that whatever similar compounds the two liquids contain that may be responsible for swelling had to be present in the lower-boiling fraction of the petroleum-derived fuel.

In an effort to determine which classes of compounds would induce the most swelling, the JP-5 fuel was distilled at specific cut-points resulting in fractions that contained different classes of organic compounds. For instance, the fraction of a JP-5 fuel with boiling point less than that of dodecane (212°C) induced considerably more swelling of nitrile rubber o-rings than the fraction with boiling point greater than 212°C (16.4 % vs. 9.2 %). Furthermore, the fraction with boiling point less than 165°C was able to swell the o-rings 30.1%. The lower boiling fraction primarily contained single-ring aromatic compounds (alkylbenzenes), a small amount of naphthalene, and alkyl phenols, all of which are suspected swelling agents.

Another attempt to identify the species contained in JP-5 that may be responsible for swelling was made by conducting switch-loading tests. The nitrile o-rings were immersed in a JP-5 fuel where they swelled. They were then transferred to a synthetic S-5 fuel where the o-rings subsequently shrunk. It was assumed that when the o-rings shrunk, they released a representative combination of the components of the JP-5 fuel that they had previously absorbed. The aromatic and oxygenate components released into the S-5 fuel were then identified using methanol extraction followed by GC/MS analysis. The most prominent oxygenates were diethylene glycol monomethyl ether (diEGME) and diethylene glycol monobutyl ether (diEGBE). Other species detected included naphthalene and two methylnaphthalene isomers. The GC/MS chromatogram for the extract is shown in Figure 1.

Both oxygenates were then tested separately for their ability to promote swelling of nitrile o-rings by spiking them at the 1% volume level in the S-5 fuel. The data in Table II indicated that both diEGME and diEGBE were relatively poor swelling promoters. Others have tested the same compounds using optical dilatometry to measure the degree of o-ring swelling and have found a higher degree of swelling (*9*). The reason for the discrepancy is not known, although the history of the nitrile rubber may play a role in its susceptibility to swelling.

The coal-derived jet fuel precursor was characterized using methanol extraction followed by GC/MS analysis of the extract to determine the most abundant aromatic compounds in the fuel. The results showed that naphthalene and naphthalene analogues were present in high concentrations. Naphthalenes were also detected in the methanol extract of S-5 fuel that was used in switch-load testing of o-rings initially immersed in JP-5 fuel (Figure 1). Therefore naphthalene and some of its saturated analogues were then tested as swelling agents at the 1% level in a synthetic S-5 jet fuel. In the progression from diaromatic to dicyclic compounds (naphthalene → tetralin → decalin) there was

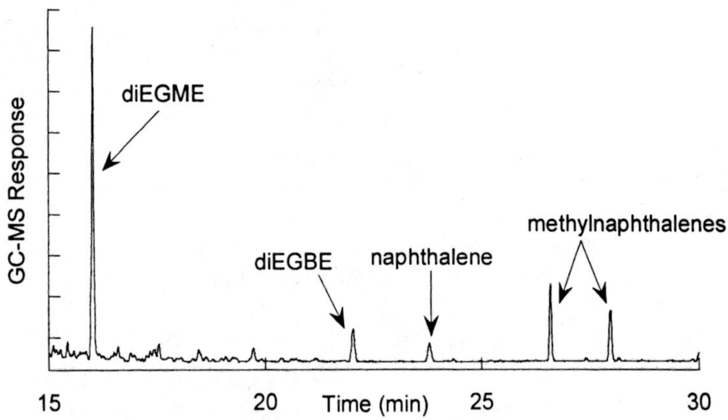


*Figure 1. GC/MS analysis of a methanol extract of S-5 fuel from switch-load testing of o-rings immersed in JP-5 followed by immersion in S-5.*

**Table II. Swelling Results for Nitrile Rubber O-Rings in F-T Jet Fuel (S-5) Spiked With Various Organic Compounds**

| *Fuel* | *Volume Change (%)* |
|---|---|
| S-5 + 1% diEGME | 1.7 ± 0.1 |
| S-5 + 0.5% diEGBE | 1.8 ± 0.3 |
| S-5 + 1% (wt) Naphthalene | 1.2 ± 0.5 |
| S-5 + 6.4% (wt) Naphthalene | 8.0 ± 0.6 |
| S-5 + 1% Tetralin | 0.9 ± 0.1 |
| S-5 + 1% Decalin | 0.8 ± 0.1 |
| S-5 + 1% Propylene Carbonate | 1.6 ± 0.1 |
| S-5 + 1% Ethylene Carbonate | 1.1 ± 0.4 |
| S-5 + 1% Vinylene Carbonate | 1.8 ± 0.4 |
| S-5 + 1% γ-butyrolactone | 2.4 ± 0.2 |
| S-5 + 1% ε-caprolactone | 4.6 ± 0.1 |
| S-5 + 1% Acetyl-Acetone | 2.6 ± 0.3 |
| S-5 + 1% Allyl Alcohol | 6.0 ± 0.3 |
| S-5 + 1% Tiglic Aldehyde | 2.3 ± 0.2 |
| S-5 + 1% Methyl Vinyl Ketone | 2.9 ± 0.3 |

NOTE: Values are reported with a 95% confidence interval.

a small decrease in the ability of the additives to enhance swelling, although none of those compounds worked particularly well at the 1% level. Naphthalene began to induce significant swelling near its solubility limit in S-5, but that concentration would be unreasonable if it were considered as an additive.

Based on the work of others (*8*), it was decided to further explore the use of Hansen Solubility Parameter (HSP) matching to predict which solvents would work best at promoting the swelling of nitrile rubber. The solubility parameters are part of an approach for predicting polymer solubility based on the theory that the cohesive energy of a liquid is comprised of several individual forces. The main forces are D (dispersion), P (dipole interactions), and H (hydrogen bonding). Maximum solubility for the polymer is predicted when the D, P and H factor values are similar for the polymer and solvent. The HSP Handbook (*10*) was scanned for solvents whose parameters most closely matched those of nitrile rubber. Compounds containing sulfur and nitrogen functionalities were ignored because they would likely negate the environmental benefits of using ultra-clean F-T fuels. Of the compounds that remained, the best HSP matches were found for oxygenates such as organic carbonates and lactones, plus select aldehydes, ketones and alcohols. Solubility issues were encountered with some of the compounds; specifically with those having a high hydrogen-bonding factor. Results are reported in Table II for the compounds that were tested. It can be plainly seen that except for the addition of allyl alcohol, there was only minor enhancement in nitrile o-ring swelling over that measured with S-5 alone.

The initial pretext for looking more closely at HSP matching to predict which compounds would be most active for o-ring swelling was the excellent swelling that was observed for nitrile rubber when phenol was added to S-5 (*9*). Those unreported results were confirmed here. The HSP for phenol is fairly close to that for nitrile rubber and therefore it was suggested that the HSP might be a good predictor of nitrile solubility (swelling). However the results reported above show that HSP matching primarily yielded candidate compounds with poor swelling efficiencies once tested. This may be because HSP factors are based on the interaction of a single solvent with a single solute. In the case of additized fuels interacting with nitrile rubber, multiple compounds are involved in solvent-solute and solvent-solvent interactions at the same time.

Because phenol yielded such favorable swelling results, along with the fact that previous analyses of JP-5 fuel indicated that the most abundant oxygenates in the fuel were alkylphenols, it was decided to examine other phenol analogues. Naphthols were included because they contain both the –OH functionality and aromatic or cyclic functionalities, both of which were suspected as being good promoters of swelling. The results are reported in Table III.

All the phenol analogues yielded enhanced swelling. The relative degree of swelling was lower for two-ring structures and when a tert-butyl group was added to the phenol. A greater degree of substitution on the ring also yielded

**Table III. Swelling Results for Nitrile Rubber O-Rings in F-T Jet Fuel (S-5) Spiked With Various Phenol Compounds**

| *Fuel* | *Volume Change (%)* |
|---|---|
| S-5 + 1% (wt) Phenol | 15.4 ± 1.0 |
| S-5 + 1% 1,2,3,4-tetrahydro-1-naphthol | 4.6 ± 0.0 |
| S-5 + 1% (wt) 5,6,7,8-tetrahydro-1-naphthol | 7.3 ± 0.4 |
| S-5 + 1% 2,4-dimethyl Phenol | 12.8 ± 0.4 |
| S-5 + 1% 2-ethyl Phenol | 13.5 ± 0.5 |
| S-5 + 1% 2-propyl Phenol | 12.0 ± 0.2 |
| S-5 + 1% 2-tert-butyl Phenol | 8.9 ± 0.1 |

NOTE: Values are reported with a 95% confidence interval.

progressively poorer results compared to the non-substituted phenol. This indicates that steric hindrance brought on by the larger double-ring structures, and multiple and larger substituents to the aromatic ring, outweigh the benefits provided hydrogen bonding and other electronic interactions between the phenol and isobutyronitrile molecules in determining the ability of the phenol to induce swelling.

Some additional tests were carried out using a combination of additives to test whether combining two additives may yield a synergistic effect that would result in greater swelling of the nitrile rubber than if either additive acted alone. Such an interaction might explain why the fraction of JP-5 that boils below 165°C is so active in inducing nitrile rubber to swell when the major non-paraffinic components of that fraction seemingly are poor swelling agents when tested individually. The binary additives that were selected for testing were combinations of the compound found to be most active for swelling, phenol, and either aromatic compounds or allyl alcohol, which was one of the most active for swelling out of all the compounds tested. The results reported in Table IV show that for the three binary combinations tested, swelling was approximately additive based on the swelling properties of the individual compounds and no synergistic promotion of swelling was observed.

A computational modeling study was initiated to determine if such modeling could predict what types of compounds would best enhance the swelling of nitrile rubber. Computational modeling of the interaction of oxygenated organic molecules with the isobutyronitrile monomer, serving as a very simplified model for the polymer, indicated that phenols have the greatest potential for o-ring swelling. For example the largest calculated interaction energy was for the 1-naphthol - isobutyronitrile system, Figure 2. For 1-naphthol, the phenolic hydrogen is preferentially attracted to the nitrogen atom of the isobutyronitrile monomer. While this is the typical interaction seen among the phenols studied, 1-naphthol also benefits from a geometrically possible interaction of the isobutyonitrile methyl group hydrogens with the aromatic pi-electron system on both rings of 1-naphthol.

**Table IV. Swelling Results for Nitrile Rubber O-Rings in F-T Jet Fuel (S-5) Spiked With Binary Additives**

| *Fuel* | *Volume Change (%)* |
|---|---|
| S-5 + 0.5% (wt) Phenol + 1% Tetralin | 7.7 ± 0.2 |
| S-5 + 0.5% (wt) Phenol + 0.5% (wt) Naphthalene | 7.7 ± 0.4 |
| S-5 + 0.5% (wt) Phenol + 1% Allyl Alcohol | 13.5 ± 0.1 |
| S-5 + 0.5% (wt) Phenol | 7.3 ± 0.4 |

NOTE: Values are reported with a 95% confidence interval.

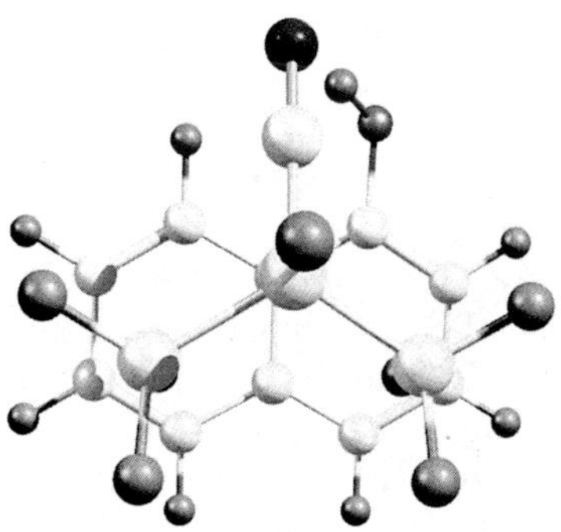

*Figure 2. Combined optimized structure 1-naphthol and isobutyronitrile.*

The compound predicted to have the second greatest potential for swelling nitrile rubber was 2-naphthol. Its optimized interaction with isobutyronitrile is shown in Figure 3. The position of the -OH group in 2-naphthol precludes the simultaneous occurrence of the phenolic hydrogen-nitrile interaction and the much weaker methyl-hydrogen aromatic pi-electron system interaction.

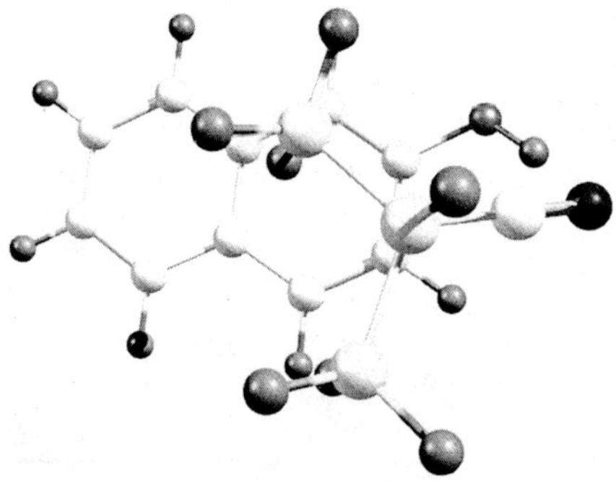

*Figure 3. Combined optimized structure 2-naphthol and isobutyronitrile.*

After the non F-T product phenols, the next strongest interaction energies were those calculated for the interactions of esters with isobutyronitrile. Figure 4 shows the optimized combined structures of the ester methylpropanoate with both of the slightly more complex rubber model monomers, cis and trans-2-methyl-hex-4-ene nitrile. In these cases the binding benefits from interactions of the nitrile group with the ester hydrogens, and the ester carboxyl group with the nitrile hydrogens. The olefin functionality has only an indirect and slightly advantageous effect toward increased binding interaction energies.

Based on the computational modeling results, additional experiments were carried out to measure the swelling of nitrile rubber o-rings in S-5 fuel additized with 1-naphthol. The first experiment was done using 0.5% naphthol because of its limited solubility in the S-5 fuel. Later, it was discovered that the solubility of 1-naphthol could be increased if 1-octanol was added as a co-additive. The swelling results are presented in Table V. It can be seen that the co-additive induced significant swelling of the nitrile rubber o-rings, but that the level of swelling was still less than that reported for phenol in Table III. The total amount of swelling induced by 1-naphthol when accompanied by the co-additive 1-octanol was approximately equal to the sum of the amounts of swelling measured when the additives were tested individually.

The last experiment that was carried out based on the computational results was one using a biodiesel B100 fuel to swell the o-rings. The biodiesel fuel is a

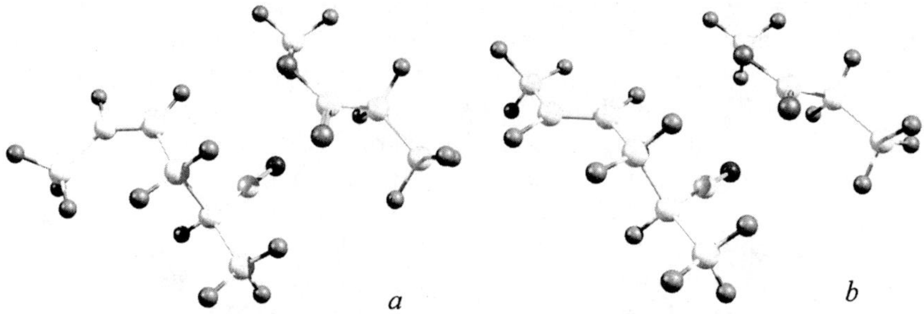

a b

*Figure 4. Combined optimized structures (a) methylpropanoate and cis 2-ethyl-hex-4-ene nitrile; (b) methylpropanoate and trans 2-methyl-hex-4-ene nitrile.*

**Table V. Swelling Results for Nitrile Rubber O-Rings in F-T Jet Fuel (S-5) Spiked With 1-Naphthol and 1-Octanol**

| *Fuel* | *Volume Change (%)* |
|---|---|
| S-5 + 0.2% (wt) 1-Naphthol | 3.3 ± 0.1 |
| S-5 + 1% 1-Octanol | 2.5 ± 0.3 |
| S-5 + 0.5% (wt) 1-Naphthol + 1% 1-Octanol | 7.8 ± 0.1 |
| S-5 + 1% (wt) 1-Naphthol + 1% 1-Octanol | 13.3 ± 0.6 |

NOTE: Values are reported with a 95% confidence interval.

mixture of mono-alkyl esters of long chain fatty acids. The o-rings were immersed directly into pure biodiesel fuel. The measured volume change was 31.8 ± 0.3%. The o-rings were then switch loaded into S-5 and that fuel was then extracted with methanol after the o-rings shrunk. GC/MS analysis of the extract showed that the major component of the biodiesel fuel absorbed by the o-rings was a C-13 methyl ester. No additional experiments were conducted using the C-13 methyl ester as a single additive to S-5 because esters are prone to undergo oxidative degradation at jet fuel operating temperatures.

## Conclusions

Through the use of switch-loading tests and an analytical method based on methanol extraction followed by GC/MS analysis, several possible compounds were identified for individual testing as possible swelling agents for synthetic fuels in contact with nitrile rubber. Those compounds included naphthalenes and alkyl phenols. Another empirical method based on matching of Hansen Solubility Parameters also was used to identify additional candidate compounds to promote swelling of synthetic fuel. Computational modeling also identified additional compounds that should promote nitrile rubber swelling. Of all the compounds tested, phenol and its analogues provide the greatest degree of swelling per volume of the compound added to the fuel. Steric factors appear to limit the ability of the phenol analogues to promote swelling. Based on the results for the compounds tested, synergistic interactions between multiple additives which would promote swelling above what would be expected from simply adding the effects of the individual components do not take place.

## Acknowledgements

The authors acknowledge the U.S. Navy for providing the JP-5 fuel and John L. Graham of the University of Dayton Research Institute for helpful discussions. Reference to any specific commercial product, process or service is to facilitate understanding and does not imply its endorsement or favoring by the U.S. Department of Energy.

## References

1. Cookson, D. J; Smith, B. E. *Fuel* **1989**, *68*, 776-781.
2. Anastopoulos, G.; Lois, E.; Zannikos, F.; Kalligeros, S.; Teas, C. *Fuel* **2002**, *81*, 1017-1024.
3. Morgan, P.; Viljoen, C.; Roets, P.; Schaberg, P.; Myburgh, I.; Botha, J.; Dancuart, L. *SAE Technical Paper No. 982488* **1998**.
4. Chang,, P. H.; Colbert, J. E.; Hardy, D. R.; Leonard, J. T. *Prepr. Pap.-Am. Chem. Soc., Div. Pet. Chem.* **2004**, *49* (4), 414-417.
5. Link, D. D.; Baltrus, J. P.; Zandhuis, P. H.; Hreha, D. C. *Energy & Fuels* **2005**, *19*, 1693-1698.
6. Alfonso, D. R., Cugini, A. V. *Prepr. Pap.-Am. Chem. Soc., Div. Fuel. Chem.* **2003**, *48* (2), 508-509.
7. Frisch, M. J. *et al.*, Gaussian 03, revision B.05; Gaussian Inc.: Wallingford CT, 2004.
8. Graham, J. L.; Striebich, R. C.; Minus, D. K.; Harrison, W. E. III *Prepr. Pap.-Am. Chem. Soc., Div. Pet. Chem.* **2004**, *49* (4), 435-439.
9. Graham, J. L., University of Dayton Research Institute, unpublished results.
10. Hansen, C. M. *Hansen Solubility Parameters, A User's Handbook*, CRC Press, Boca Raton, FL, 2000.

# Process Modeling and Optimization

# Chapter 16

# Artificial Neural Network-Aided Catalyst Research for Low-Pressure DME Synthesis from Syngas

**Kohji Omata, Sutarto, Masahiko Hashimoto, Gunji Ishiguro, Yuhsuke Watanabe, Tetsuo Umegaki, and Muneyoshi Yamada**

**Department of Applied Chemistry, Graduate School of Engineering, Tohoku University, Aoba 6–6–07, Aramaki, Aoba-ku, Sendai 980–8579, Japan**

Artificial neural network (ANN) was applied to screen additives of Cu/Zn based catalyst which is effective for high activity of methanol synthesis from syngas in one-step DME synthesis at low pressure (1-3 MPa) and low temperature (500-550K). Besides the conventional additives such as Al, Ti and V were discovered to be effective additives for high activity of methanol synthesis when Cu/Zn based catalyst were prepared by oxalate-ethanol co-precipitation method and mixed with γ-alumina. They were concluded to mitigate the inhibiting effect of $H_2O$ formed at high CO conversion. ANN was also applied for optimization of catalyst composition. The methodology comprising of DOE, ANN and grid search, was confirmed to be useful and can cope with the catalyst optimization tailored for each process situation.

## Introduction

Dimethyl ether (DME), the target product of this study, has many advantages when it is used as diesel fuel: sulfur-free, low particulate emission and high cetane number. So, it is one of the superior candidates for high-quality diesel fuel of next generation, and the world first industrial DME plant for LPG alternative fuel started-up in 2003 (*1*). Besides the large scale process, we proposed compact and economic process in order to produce DME from dispersed unused carbon resources (*2*). To reduce much production costs of DME in a small-scale process, low pressure-operation and high one-pass conversion are required. More specifically, operation under syngas pressure at the outlet of reforming reactor (i.e. 5 MPa, or hopefully 1-3 MPa) is desirable in order to omit huge booster compressor, and as for the conversion, over 90% should be required because recycling loop of unreacted gas can be omitted.

Because methanol synthesis from syngas (eq.1) is exothermic reaction, operation at low temperature (400 - 480 K) is required for 90% CO conversion at 1-3 MPa.

$$CO + 2H_2 \rightarrow MeOH \quad (1)$$

Low-temperature methanol synthesis process was proposed for such high CO conversion. In the process, nickel and alkoxide catalyst was applied in liquid phase at 373 K, 1MPa. 90% CO conversion was reported at the mild conditions (*3, 4*). Unfortunately the alkoxide is not stable with water and carbon dioxide and thorough removal of the impurity in syngas is necessary to avoid deactivation. Industrialization of the process is not realized by the defect of the catalyst.

On the other hands, operation at 500-550 K is possible for 90% CO conversion at 1-3 MPa in one-step DME synthesis from syngas (eq.2) where hybrid catalyst with Cu-Zn based oxide and solid acid is used, because the equilibrium conversion is higher than that of methanol synthesis (eq.1).

$$3CO + 3H_2 \rightarrow DME + CO_2 \quad (2)$$

At these temperatures, not alkoxide catalyst, but conventional Cu/Zn based catalyst could be applied for the process to attain such high CO conversion, because some additives to Cu/Zn based catalyst such as Al, Ga, Zr, and Cr optimized the surface $Cu^0/Cu^+$ ratio to give high methanol synthesis activity (*5*). Furthermore, Cu/Zn based catalysts prepared by an oxalate-ethanol co-precipitation method shows higher activity than those prepared by alkali co-precipitation method (*6, 7*), and thus, novel synergetic additives are expected using the novel preparation method. Therefore, we focused on the development of the active Cu-Zn based catalyst prepared by oxalate-ethanol co-precipitation method, and screened effective additives for high activity of one-step DME synthesis from syngas.

Combinatorial approach attracts much attention as a successful tool not only for material development but also for heterogeneous catalyst development (*5-11*). The approach seems to be effective to optimize the catalyst in a short period. Among them, however, high throughput screening (HTS) tool available at high pressure reaction is very few (*12-13*). Recently, ANN has been applied to catalyst development through the prediction of catalyst characteristics such as activity, selectivity and durability, based on experimental results. Prediction of the characteristics of new catalysts or catalyst additives from the physicochemical properties of elements would accelerate catalyst development (*14, 15*). We recently succeeded in discovering an effective additive for Ni/active carbon catalyst for the carbonylation of methanol based on previous experimental results and the physicochemical properties of the additive elements (*16*). We applied ANN to both the screening and the optimization of Cu/Zn based catalyst in this study. The combination of ANN and conventional high pressure reactor functioned as high pressure HTS reactor. We expect that this methodology will contribute to rapid catalyst development.

## Experimental

### Catalyst Preparation

Cu/Zn/X oxide catalysts were prepared by an oxalate-ethanol co-precipitation method. Ethanol solutions of nitrate of Cu, Zn and one additive were mixed (Cu/Zn/X molar ratio was 65/35/5), and then an ethanol solution of oxalic acid was mixed to precipitate the mixed oxalic salts. Only vanadium acetyl acetone was solved in a mixed solvent (acetone:methanol =1:1), and it was used as V source. Ethanol was removed by vaporization followed by calcination in air at 623 K for 4 hours.

### Activity Test

One-step DME synthesis from syngas was conducted. The Cu/Zn based catalyst was mixed with γ-alumina granule. The weight ratio of the alumina was 1.2 times to that of the Cu/Zn based catalyst. The hybrid catalyst was activated in-situ in the conventional fixed bed reactor at 0.1 MPa and 523 K using reaction gas ($H_2/CO/CO_2/N_2$ = 60/30/5/5). DME synthesis was conducted at 498 K, 1 MPa, W/F= 8 g-cat. h/mol to measure COx (CO and $CO_2$) conversion. The product was analyzed by micro gas chromatograph (Micro-GC, M-200, Agilent Technologies, Inc.). In the case of methanol synthesis from syngas, the Cu/Zn based catalyst was used at 498 K, 1-3 MPa, W/F= 5-10 g-cat. h/mol. In some cases, water was added to the syngas by a cylinder pump at 0-20 mmol/g-cat./h. For optimization of Cu/Zn based catalyst, the composition was designed by L18 orthogonal array (*17*) and the activity test was conducted in 51 line HTS reactor similar to 96 line HTS reactor (*13, 18*). The reaction condition was: 509-548K, 1 MPa, $H_2/CO/CO_2/N_2$ = 44/22/30/4. The activity of each catalyst was determined using color change of chromium ion in water by the trapped methanol as reduction reagent (*13, 18*).

### Methodology

Artificial neural networks were applied to correlate physicochemical properties of additive X with the activity of Cu/Zn/X catalyst. Among many types of ANN, radial basis function network (RBFN) was used because it is robust and experimental error-tolerant. STATISTICA Neural Network (version 6; StatSoft, Inc.) was used for constructing RBFN. Eight elements (B, Re, Nb, Cd, Ce, K, Sm, Tl) were selected as additives (X), and Cu/Zn/X was prepared by ethanol co-precipitation method and tested. Physicochemical property of these elements collected from database software (Periodic Table X)(*19*) were used as input data of RBFN, and experimentally determined activity of Cu/Zn/X catalysts were used as output data for training. After training the RBFN predicted the effective additives for high activity based on the physicochemical property of the additives. In the case of the optimization of catalyst composition, ANN was again used to correlate catalyst composition with the activity. Orthogonal array was used to design catalyst composition. These catalysts were prepared and tested. The catalyst composition was used as input data of RBFN, and the catalytic activity was used as output data for training. Then, grid search was conducted to find the global maximum of the ANN. The grid search program was coded using macro commands of STATISTICA.

## Result and Discussion

### Screening of Effective Additives for One-Step DME Synthesis from Syngas

In the first step, ANN was applied to screen the most effective additive X for high activity based on both the physicochemical properties of element X and the activity of Cu/Zn/X catalyst for one-step DME as experimentally determined (*20*). The range and dispersion of physicochemical properties included in the training data were important factors for the precision of the predictions (*16, 21*). In the present study, therefore, 8 elements (B, Re, Nb, Cd, Ce, K, Sm, Tl) were selected with dispersed properties. The physicochemical properties employed were: heat of vaporization (HV, kJ/mol), photoelectric work function (PW, eV), covalent radius (CR, pm), ionic radii (IR, pm), first ionization energy (1I, eV), second ionization energy (2I, eV), density (DS, g/cm$^3$), thermal conductivity (TC, W/mK) with atomic number (AN).

After catalytic activity was measured at 1 MPa, 498K, the activity change with time on stream was fitted to a generalized power-law equation (GPLE) (*22*). The activity was expressed as: COx conversion = a / ( b + t ) ^ c + d where t (min) is time on stream. Each catalyst was featured by the four parameters, and information mining was carried out from the unsteady catalytic activity. This kind of parameterization was necessary because a variety of additives were used to obtain training data of ANN, and unsteady activity was sometimes obtained. The properties of additives and experimental results are summarized in Table I. A radial basis function network (RBFN), a type of ANN, was employed to correlate the physicochemical properties of the additive element to the activity. The experimental result of Cu/Zn/Al was used as validation data. After training, validation error of RBFN7-8-4 with one hidden layer was the lowest among various kind of RBFNs. Number of nodes in the input layer, hidden layer, and output layer of RBFN7-8-4 is 7, 8, and 4, respectively. Seven properties (i.e. AN, HV, CR, DS, PW, 1I, 2I) were used for input layer, and IR and TC were not included in the RBFN. Then, the periodic table was surveyed to discover the effective additives. Using the trained RBFN and the predicted GPLE parameters, COx conversion at 500 min time-on-stream was calculated. Cu/Zn/V and Cu/Zn/Ti were predicted to be active as "predicted" in Table I. Although the experimentally decided activities of both were lower than that of validation data (Cu/Zn/Al), and Cu/Zn/Ti showed rather higher activity than predicted as "experimental" in Table I, the trained RBFN succeeded in discovering new candidates of effective additive.

For methanol synthesis from syngas, B and Sc were discovered by conventional method as effective additive of Cu/Zn based catalyst (*23*). On the contrary, neither Ti nor V was discovered by the method, and both are not well-known as additives of Cu/Zn based catalyst for methanol synthesis. Because the reaction conditions in both cases are similar (1 MPa, 498K), it suggests that Ti and V are specific for DME synthesis. Considering the circumstance of catalyst under reaction, our attention was paid to $H_2O$ formed by dehydration of methanol during the one-step DME synthesis because we already reported that addition of $H_2O$ retards the activity of methanol formation by Cu/Zn based catalyst in a reversible manner (*13*). Thus it was suggested that Ti and V probably mitigate the inhibiting $H_2O$ effect. Effect of $H_2O$ addition during methanol synthesis from syngas at 1MPa, 498K was ascertained as shown in

Figure 1. The activity of Cu/Zn/Al decreased as the partial pressure of $H_2O$ increased, but the decrease of the activity of Cu/Zn/Ti was gradual. More remarkable difference was observed in Cu/Zn/V. While the activity of Cu/Zn/V is the lowest among the three catalysts without $H_2O$ addition, the activity of Cu/Zn/V is much higher under high $H_2O$ partial pressure. Clearly V gives high tolerance against water to Cu/Zn based catalyst.

The result suggests that Cu/Zn/V exert its potential activity at high CO conversion in one-step DME synthesis because partial pressure of $H_2O$ is high at high CO conversion. The hypothesis obtained by reaction using $H_2O$, was tested with one-step DME synthesis under 3 MPa to obtain high CO conversion. As shown in Figure 2, the activities of Cu/Zn/Ti and Cu/Zn/V for one-step DME synthesis were almost same when they were used alone. On the other hand in 2-layers experiment where Cu/Zn/Ti was placed in the upper part of the catalyst bed and Cu/Zn/V was placed in the lower part, CO conversion was much higher although total W/F was identical. CO conversion was probably improved by Cu/Zn/V catalyst at the bottom of reactor.

The circumstance of catalyst bed is usually varied from top to bottom of the bed when total conversion is high, and consequently, desired character of catalyst is also varied. In one-step DME synthesis, Cu/Zn based catalyst seems to be sensitive to the circumstance from this result, and it was suggested that optimization is required to correspond with the variety of reaction conditions in order to achieve 90% CO conversion at milder conditions. In the next section, the methodology will be unveiled to predict the optimum catalyst composition for different reaction temperature and will be checked if the predicted optimum composition is varied reasonably corresponding with the reaction temperature.

## Optimization of Catalyst Composition of Cu/Zn Based Catalyst for One-Step DME Synthesis from Syngas.

Combinatorial approach attracts much attention as a successful tool for heterogeneous catalyst development. We first developed high pressure HTS reactor with 12 lines as a combinatorial tool (*24*). Home-made genetic algorithm (GA) program was then used with HTS reactor to find the optimum catalyst composition (*25, 26*). In our GA program, gene, fitness and environment were catalyst (presented by binary code), activity, and reaction conditions, respectively. In the next stage, an artificial neural network (ANN) was introduced for activity mapping (*18, 27-30*) in order to reduce laborious experimental steps to evaluate the fitness of gene. After ANN is trained by the initial experimental results, it can evaluate the fitness of gene in a GA program instead of experiments at every generation. Many successful results in solid

**Table I. Physicochemical Property of Additive X and the GPLE Parameters of Cu/Zn/X Catalyst**

| catalyst | ANN input data (physicochemical property) | | | | | | | | | ANN output (GPLE parameter) | | | | calculated catalyst activity at 500 min (%) |
|---|---|---|---|---|---|---|---|---|---|---|---|---|---|---|
| | AN | HV (kJ/mol) | PW (eV) | CR (pm) | IR (pm) | 1I (eV) | 2I (eV) | DS (g/cm³) | TC (W/mK) | a | b | c | d | |
| training data | | | | | | | | | | | | | | |
| Cu/Zn/B | 5 | 504.4 | 4.45 | 81 | 23 | 8.298 | 25.15 | 2.34 | 27 | 4.79 | 5.27 | 0.17 | 0.43 | 2.1 |
| Cu/Zn/Re | 75 | 704.0 | 4.72 | 128 | 72 | 7.833 | 13.06 | 21.00 | 47.9 | 0.85 | 11.66 | 0.24 | 0.06 | 0.3 |
| Cu/Zn/Nb | 41 | 680.1 | 4.33 | 134 | 74 | 6.758 | 14.32 | 8.55 | 53.7 | 8.29 | 79.57 | 0.11 | 3.25 | 7.4 |
| Cu/Zn/Cd | 48 | 100.0 | 4.08 | 141 | 114 | 8.993 | 16.91 | 8.65 | 96.8 | 18.3 | 13.91 | 1.29 | 0.00 | 0.0 |
| Cu/Zn/Ce | 58 | 398.0 | 2.90 | 165 | 107 | 5.538 | 10.85 | 6.78 | 11.4 | 10.4 | 20.67 | 0.21 | 1.54 | 4.3 |
| Cu/Zn/K | 19 | 79.1 | 2.29 | 203 | 138 | 4.340 | 31.63 | 0.86 | 102.4 | 9.93 | 42.51 | 0.99 | 0.22 | 0.2 |
| Cu/Zn/Sm | 62 | 164.8 | 2.70 | 166 | 111 | 5.643 | 11.07 | 7.54 | 13.3 | 9.44 | 20.72 | 0.16 | 2.33 | 5.8 |
| Cu/Zn/Tl | 81 | 166.0 | 3.84 | 155 | 149 | 6.108 | 20.43 | 11.85 | 46.1 | 1.05 | 8.68 | 0.50 | 0.01 | 0.1 |
| validation data | | | | | | | | | | | | | | |
| Cu/Zn/Al | 13 | 290.8 | 4.17 | 125 | 54 | 5.985 | 18.82 | 2.70 | 273.0 | 10.71 | 60.06 | 0.03 | 3.66 | 12.5 |
| physicochemical property for prediction and predicted GPLE parameter | | | | | | | | | | | | | | |
| Cu/Zn/V | 23 | 459.7 | 4.30 | 122 | 72 | 6.746 | 14.66 | 5.80 | 30.7 | 11.29 | 74.11 | 0.08 | 3.84 | 10.6 |
| Cu/Zn/Ti | 22 | 425.5 | 4.33 | 132 | 80 | 6.828 | 13.58 | 4.50 | 21.9 | 12.59 | 70.74 | 0.16 | 3.67 | 8.2 |
| experimental GPLE parameter | | | | | | | | | | | | | | |
| Cu/Zn/V | - | - | - | - | - | - | - | - | - | 13.61 | 26.10 | 0.10 | 3.19 | 10.7 |
| Cu/Zn/Ti | - | - | - | - | - | - | - | - | - | 40.88 | 88.60 | 0.48 | 10.09 | 12.1 |

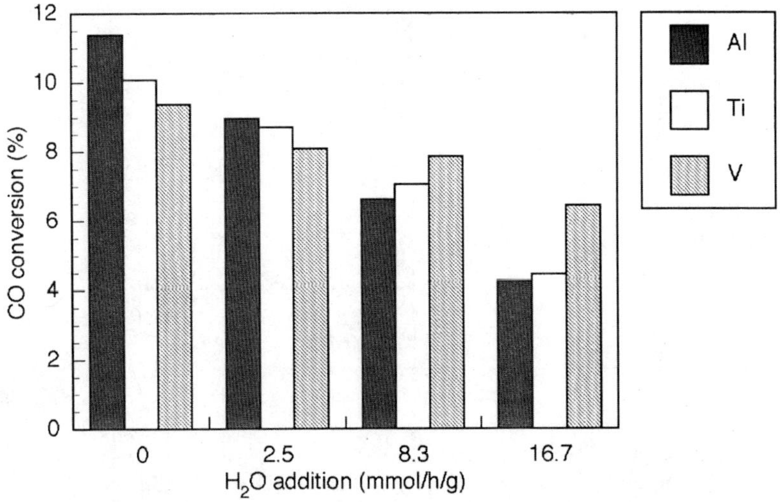


*Figure 1. Effect of Al, Ti and V addition on MeOH synthesis at 1 MPa, 498K. W/F=8 gh/mol (dry base).*

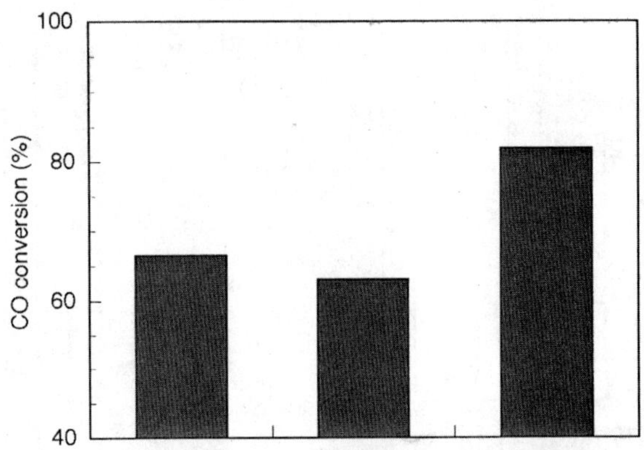


*Figure 2. Effect of layered catalyst bed at 3 MPa, 498K. W/F=10 gh/mol. 0.8g Cu/Zn based catalyst and 1.2g $\gamma$-$Al_2O_3$ were mixed. 2-layers: Upper(Cu/Zn/Ti)=0.6 g and 0.9 g $\gamma$-$Al_2O_3$, lower(Cu/Zn/V)=0.2 g, 0.3g $\gamma$-$Al_2O_3$*

catalyst development using similar methodology (HTS + ANN + GA), were also reported and reviewed (*31-33*).

One possible bottle neck of our optimization method was the training of ANN. Random parameters for catalyst were used in these our early experiments to obtain training data for ANN because such dispersed data were thought to give better ANNs. However, rather large numbers of dataset were necessary in this method, and additional data were sometimes required for re-training of ANN for more precise prediction. A better and well-designed dataset for effective training of ANN was necessary. The statistical design of experiment (DOE) was applied to find important factors among catalyst composition, preparation parameters and reaction conditions (*34-37*). The quantitative influence of the parameter was often estimated using a polynomial model and a regression technique (*36*). Especially the orthogonal array can reduce the number of necessary experiments, while maximizing the amount of information derived from the reduced experiment set. Such arranged datasets seems suitable for ANN training (*36-40*) and the resulting ANN gave good predictions. The more precise prediction is of cause possible by ANN trained by large amount of dataset when HTS is available. Such integrated methodology (*41*) is attractive for rapid catalyst development.

Another drawback was a criterion of convergence by GA. It is intrinsically difficult or impossible to verify the goal of optimization by GA. In fact fluctuation of maximum value and that of dispersion of catalyst composition are main indexes for the convergence. Instead of GA, therefore, more straightforward method was applied. It is so-called "grid search" (GS): all the activities of all possible combination of catalyst component were predicted by ANN. The global optimum on the ANN was found rapidly with the assistance of GS (*13, 23, 39, 40, 42- 44*).

The finally obtained methodology (DOE + HTS + ANN + GS) was used to find the optimum catalyst compositions of Cu/Zn based catalyst at different reaction temperature. In Table II experimental design using a part of $L_{18}$ orthogonal array is summarized. For each factor (catalyst element) three level of relative amount was set up as shown at the bottom of the table, and catalyst composition was determined according to the array. For example in No.8 experiment, Cu:Zn:Al:Ti:Nb:V:Cr = 3:2:3:2:1:3:1 (level) = 120:30:40:15:0:30:0 (molar ratio) = 51.1:12.8:17.0:6.4:0.0:12.8:0.0 (mol%). The catalyst with this composition was prepared by oxalate-ethanol method and the activity test was conducted in HTS reactor in Figure 3, where product MeOH was guided through capillary tube to water trap and the each amount was determined by color reaction (*13*). Activity was indicated as space time yield (STY; g-MeOH/kg-cat./h) in Table II for each reaction temperature. Because the information on relative activity was important for optimization of catalyst composition, STY data at 519K was used without correction although it was rather lower than expected. Total 18 experimental results were used for the training of RBFN7-18-1 for each reaction temperature, and the resulting 5 RBFNs were used for grid search in order to find the optimum catalyst composition. The compositions are illustrated in Fig. 4. V was not included in the optimum composition probably because of the low CO conversion in the HTS reactor. Clear tendency, including the result at 519K, that Cu content is decreased at higher temperature was observed, and instead, the amount of additives such as Ti and Nb was increased.

Such a stimulating result was obtained using only 5 experimental HTS runs. However, as shown symbolically in V content, the reaction conditions of HTS to accumulate the training data of ANN are very important: ANN will learn exactly the result under the reaction conditions. If catalyst for high CO conversion were necessary, experiments should have been done under such a reaction conditions to give high CO conversion.

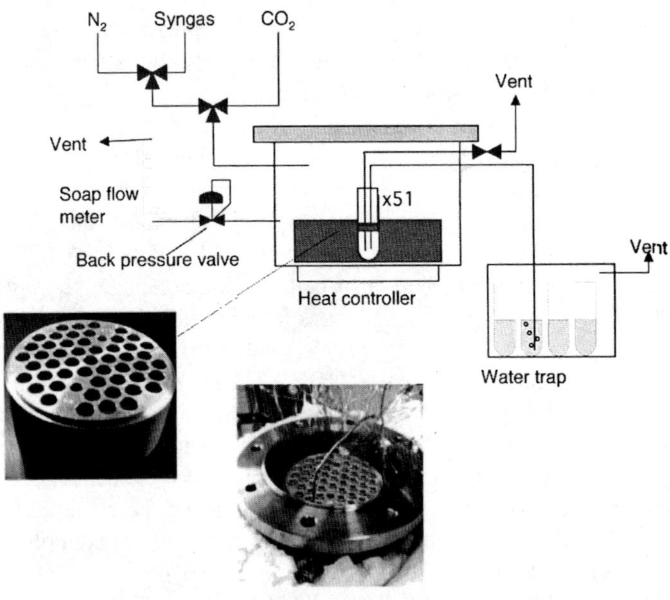


*Figure 3. High pressure HTS reactor (51 lines).*

## Conclusions

Artificial neural network (ANN) was applied to screen effective additives of Cu/Zn based catalyst with γ-alumina for high activity in one-step DME synthesis from syngas at 1 MPa, 498K. Physicochemical properties of additive X were

**Table II. Design of Catalyst Compostion and Activity for Methanol Synthesis at 1 MPa**

| | part of $L_{18}$ orthogonal array | | | | | | | catalyst composition (mol%) | | | | | | | activity (g-MeOH/kg-cat./h) | | | | |
|---|---|---|---|---|---|---|---|---|---|---|---|---|---|---|---|---|---|---|---|
| | Cu | Zn | Al | Ti | Nb | V | Cr | Cu | Zn | Al | Ti | Nb | V | Cr | 548K | 530K | 519K | 514K | 509K |
| No.1 | 1 | 1 | 1 | 1 | 1 | 1 | 1 | 61.5 | 23.1 | 15.4 | 0.0 | 0.0 | 0.0 | 0.0 | 95.2 | 170.1 | 206.3 | 210.5 | 178.3 |
| 2 | 1 | 2 | 2 | 2 | 2 | 2 | 2 | 25.8 | 19.4 | 16.1 | 9.7 | 9.7 | 9.7 | 9.7 | 58.5 | 48.9 | 35.0 | 39.8 | 31.1 |
| 3 | 1 | 3 | 3 | 3 | 3 | 3 | 3 | 16.3 | 18.4 | 16.3 | 12.2 | 12.2 | 12.2 | 12.2 | 28.4 | 31.7 | 28.2 | 22.8 | 17.0 |
| 4 | 2 | 1 | 1 | 2 | 2 | 3 | 3 | 41.0 | 7.7 | 5.1 | 7.7 | 7.7 | 15.4 | 15.4 | 40.9 | 46.5 | 34.4 | 37.4 | 30.2 |
| 5 | 2 | 2 | 2 | 3 | 3 | 1 | 1 | 41.0 | 15.4 | 12.8 | 15.4 | 15.4 | 0.0 | 0.0 | 83.0 | 116.9 | 97.7 | 113.8 | 105.4 |
| 6 | 2 | 3 | 3 | 1 | 1 | 2 | 2 | 41.0 | 23.1 | 20.5 | 0.0 | 0.0 | 7.7 | 7.7 | 72.4 | 120.9 | 72.9 | 72.3 | 79.4 |
| 7 | 3 | 1 | 2 | 1 | 3 | 2 | 3 | 51.1 | 6.4 | 10.6 | 0.0 | 12.8 | 6.4 | 12.8 | 50.1 | 49.1 | 39.1 | 48.5 | 29.3 |
| 8 | 3 | 2 | 3 | 2 | 1 | 3 | 1 | 51.1 | 12.8 | 17.0 | 6.4 | 0.0 | 12.8 | 0.0 | 66.7 | 89.1 | 65.2 | 63.8 | 61.2 |
| 9 | 3 | 3 | 1 | 3 | 2 | 1 | 2 | 51.1 | 19.1 | 4.3 | 12.8 | 6.4 | 0.0 | 6.4 | 94.8 | 130.4 | 126.3 | 150.6 | 94.4 |
| 10 | 1 | 1 | 3 | 3 | 2 | 2 | 1 | 25.8 | 9.7 | 25.8 | 19.4 | 9.7 | 9.7 | 0.0 | 26.5 | 38.9 | 20.9 | 20.6 | 23.7 |
| 11 | 1 | 2 | 1 | 1 | 3 | 3 | 2 | 25.8 | 19.4 | 6.5 | 0.0 | 19.4 | 19.4 | 9.7 | 26.4 | 39.0 | 30.6 | 33.5 | 23.3 |
| 12 | 1 | 3 | 2 | 2 | 1 | 1 | 3 | 25.8 | 29.0 | 16.1 | 9.7 | 0.0 | 0.0 | 19.4 | 65.8 | 102.8 | 48.8 | 52.7 | 70.6 |
| 13 | 2 | 1 | 2 | 3 | 1 | 3 | 2 | 41.0 | 7.7 | 12.8 | 15.4 | 0.0 | 15.4 | 7.7 | 44.1 | 49.3 | 38.4 | 36.6 | 36.8 |
| 14 | 2 | 2 | 3 | 1 | 2 | 1 | 3 | 41.0 | 15.4 | 20.5 | 0.0 | 7.7 | 0.0 | 15.4 | 58.6 | 62.6 | 42.3 | 47.6 | 32.9 |
| 15 | 2 | 3 | 1 | 2 | 3 | 2 | 1 | 41.0 | 23.1 | 5.1 | 7.7 | 15.4 | 7.7 | 0.0 | 85.1 | 120.0 | 105.5 | 115.8 | 103.1 |
| 16 | 3 | 1 | 3 | 2 | 3 | 1 | 2 | 51.1 | 6.4 | 17.0 | 6.4 | 12.8 | 0.0 | 6.4 | 54.9 | 54.8 | 43.9 | 43.7 | 30.5 |
| 17 | 3 | 2 | 1 | 3 | 1 | 2 | 3 | 51.1 | 12.8 | 4.3 | 12.8 | 0.0 | 6.4 | 12.8 | 68.4 | 70.7 | 60.2 | 62.9 | 42.7 |
| 18 | 3 | 3 | 2 | 1 | 2 | 3 | 1 | 51.1 | 19.1 | 10.6 | 0.0 | 6.4 | 12.8 | 0.0 | 76.0 | 132.1 | 117.7 | 105.5 | 104.0 |

| | Cu | Zn | Al | Ti | Nb | V | Cr |
|---|---|---|---|---|---|---|---|
| level 1 | 40 | 15 | 10 | 0 | 0 | 0 | 0 |
| level 2 | 80 | 30 | 25 | 15 | 15 | 15 | 15 |
| level 3 | 120 | 45 | 40 | 30 | 30 | 30 | 30 |

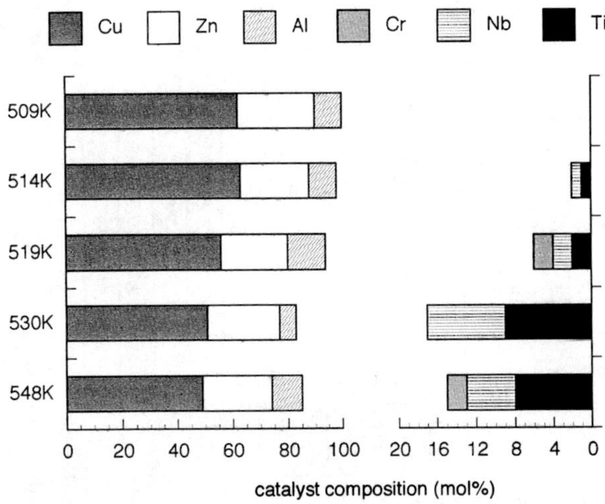


*Figure 4. Predicted optimum catalyst composition for each temperature at 1MPa.*

correlated with the activity of Cu/Zn/X catalyst through ANN training, and the trained ANN predicted that Ti and V are effective. Experimentally Cu/Zn/V showed the highest activity for MeOH synthesis from syngas containing large amount of $H_2O$. These additives were concluded to mitigate the inhibiting effect of $H_2O$ which will be pronounced at high CO conversion in one-step DME synthesis. ANN was also applied for optimization of catalyst composition. The methodology comprising of design of experiment, HTS, ANN, and grid search was confirmed to cope quite rapidly and sensitively with the catalyst optimization tailored for each process situation.

## Acknowledgment

We acknowledge financial support from Research for the Future Program of Japan Society for the Promotion of Science under the Project Synthesis of Ecological High Quality Transportation Fuels (JSPS-RFTF98P01001)

## References

1. Toyo Engineering, Press Release, Chiba, Japan, 30 January 2004.
2. Yamada, M. *Energy & Fuels* **2003**, *17*, 797.
3. Marchionna, M.;Basini, L.; Aragno, A.; Lami, M.; Ancillotti, F. *J. Mol. Catal.* **1992**, *75*, 147.

4. Palekar, V.M.; Jung, H.; Tierney, J.W. , *Appl. Catal. A* **1993**, *102,* 13.
5. Saito, M., Fujitani, T., Takeuchi, M., Watanabe, T. *Appl. Catal. A*, **1996**, *138,* 311.
6. Sun, Q.; Zhang, Y.; Chen, H.; Deng, J.; Wu, S. *J. Catal,* **1997**, *167,* 92.
7. Omata, K, Watanabe Y., Umegaki, T., Ishiguro G., Yamada, M. *Fuel,* **2002**, *81,* 1605.
8. Senkan, S. *Angew. Chem. Int. Ed.,* **2001**, *40,* 312.
9. Jandeleit, B.; Schaefer, D. J.; Powers, T. S.; Turner, H. W.; Weinberg, W. H. *Angew. Chem. Int. Ed.,* **1999**, *38*, 2494.
10. Hagemeyer, A.; Jandeleit, B.; Liu, Y.; Poojary, D. M.; Turner, H. W.; Volpe Jr., A. F.; Weinberg, W. H.; *Appl. Catal. A,* **2001**, *221,* 23.
11. Scheidtmann, J.; Weiss, P. A.; Maier, W. F. *Appl. Catal. A,* **2001**, *222,* 79.
12. Kiener, C.; Kurtz, M.; Wilmer, H.; Hoffmann, C.; Schmidt, H.-W.; Grunwaldt, J. -D.; Muhler, M.; Schüth F., *J. Catal.,* **2003**, *216,* 110.
13. Omata, K.; Hashimoto, M.; Watanabe, Y.; Umegaki, T.; Wagatsuma, S.; Ishiguro, G.; Yamada, M. *Appl. Catal. A,* **2004**, *262,* 207.
14. Kito, S.; Hattori, T.; Murakami, Y. *Ind. Eng. Chem. Res.,* **1992**, *31,* 979.
15. Kito, S.; Hattori, T.; Murakami, Y. *Appl. Catal. A,* **1994**, *114,* L173.
16. Omata, K.; Yamada, M. *Ind. Eng. Chem. Res.,* **2004**, *43,* 6622.
17. Omata, K.; Nukui, N.; Yamada, M. *Ind. Eng. Chem. Res.,* **2005**, *44,* 296.
18. Watanabe, Y.; Umegaki, T.; Hashimoto, M.; Omata, K.; Yamada, M. *Catal. Today,* **2004**, *89,* 455.
19. Periodic Table X is downloadable from the site of Synergy Creations$^{TM}$ (http:// www.synergycreations.com/).
20. Omata, K.; Sutarto; Hashimoto, M.; Ishiguro, G.; Watanabe, Y.; Umegaki, T.; Yamada, M. *Ind. Eng. Chem. Res., accepted.*
21. Omata, K; Hashimoto, M.; Sutarto; Watanabe, Y; Umegaki, T; Yamada, M. *J. Jpn. Pet. Inst.,* **2005**, *48,* 145.
22. Bartholomew, C. H. *Appl. Catal. A*, **2001**, *212,* 17.
23. Omata, K.; Watanabe, Y.; Hashimoto, M.; Umegaki, T.; Yamada, M. *Ind. Eng. Chem. Res.,* **2004**, *43,* 3282.
24. Omata, K.; Ishigure, G.; Yamada, M. *J. Jpn. Petrol. Inst.,* **2000**, *43,* 317.
25. Omata, K.; Umegaki, T.; Ishiguro, G.; Yamada, M. *Sekiyu Gakkaishi (J. Jpn. Petrol. Inst.)*, **2001**, *44,* 327.
26. Umegaki, T.; Omata, K.; Ishiguro, G.; Watanabe, Y.; Yamada, M. *J. Jpn. Petrol. Inst.,* **2003**, *46,* 181.
27. Omata, K.; Umegaki, T.; Watanabe, Y.; Yamada, M. *J. Jpn. Petrol. Inst.,* **2002**, *45,* 192.
28. Omata, K.; Umegaki, T.; Watanabe, Y.; Nukui, N.; Yamada, M. *J. Jpn. Petrol. Inst.,* **2003**, *46,* 189.
29. Omata, K.; Watanabe, Y.; Umegaki, T.; Hashimoto, M.; Yamada, M., *J. Jpn. Petrol. Inst.*, **2003**, *46,* 328.
30. Umegaki, T.; Watanabe, Y.; Nukui, N.; Omata, K.; Yamada, M. *Energy & Fuels*, **2003**, *17,* 850.

31. Corma, A.; Serra, J.M.; Argente, E.; Botti, V.; Valero, S. *Chem. Phys. Chem.* **2002**, *3*, 939.
32. Baumes, L.; Farrusseng, D.; Lengliz, M.; Mirodatos, C. *QSAR Combi. Sci.* **2004**, *23*, 767.
33. Rodemerck, U.; Baerns, M.; Holena, M.; Wolf, D. *Appl. Surf. Sci.,* **2004**, *223,* 168.
34. Baker, J. E.; Burch, R.; Golunski, S. E. *Appl. Catal. A,* **1989**, *53,* 279.
35. Dawson, E. A.; Barnes, P. A. *Appl. Catal. A,* **1992**, *90,* 217.
36. Lima, A. A. G.; Nele, M. N.; Moreno, E. L.; Andrade, H. M. C. *Appl. Catal. A,* **1998**, *171,* 31.
37. Ozdemir, C.; Nilgun, A.; Yildirim, R. *Appl. Catal. A,* **2004**, *258,* 145.
38. Huang, K.; Chen, F. -Q.; Lu, D. -W. *Appl. Catal. A,* **2001**, *219,* 61.
39. Kompany-Zareh, M.; Tavallali, H.; Sajjadi, M. *Anal. Chim. Acta,* **2002**, *469,* 303.
40. Omata, K.; Watanabe, Y.; Hashimoto, M.; Umegaki, T.; Yamada, M. *J. Jpn. Petrol. Inst.,* **2003**, *46,* 387.
41. Ohrenberg, A.; von Torne, C.; Schuppert, A.; Knab, B. *QSAR Combi. Sci.* **2005**, *24*, 29.
42. Omata, K.; Hashimoto, M.; Watanabe, Y.; Umegaki, T.; Yamada, M. *J. Jpn. Petrol. Inst.,* **2003**, *46,* 383.
43. Umegaki, T.; Watanabe, Y.; Nukui, N.; Omata, K.; Yamada, M. *Trans. Material Res. Soc. Jpn,* **2004**, *29,* 289.
44. Nascimento, C.A.O.; Giudici, R. *Comp. Chem. Eng.,* **1998**, *22,* S595.

## Chapter 17

# Slurry Bubble Column Reactor Optimization

**Isaac K. Gamwo[1], Dimitri Gidaspow[2], and Jonghwun Jung[3]**

[1]Computational Science Division, National Energy Technology Laboratory, U.S. Department of Energy, P.O. Box 10940, Pittsburgh, PA 15236
[2]Department of Chemical and Environmental Engineering, Illinois Institute of Technology, Chicago, IL 60616
[3]Mathematics and Computer Science Division, Argonne National Laboratory, 9700 South Cass Avenue, Argonne, IL 60439

Slurry bubble column reactors (SBCR) are the preferred contactors for the conversion of syngas to fuels and chemicals partially due to their superior heat and mass transfer characteristics. The multiphase fluid dynamics in these systems greatly affect the reactor volumetric productivity. Here, we have developed a computational fluid dynamics (CFD) - assisted design methodology for searching the optimum particle size for maximum production in a SBCR. Reactor optimization due to heat exchanger configuration was also investigated. We have rearranged the heat exchangers in a SBCR and constructed a CFD model for a baffled reactor. The novel arrangement of the exchangers prevents the unfavorable high catalysts concentration at the lower stage of the reactor. Thus an optimum catalyst concentration is maintained during the course of the production of liquid fuels.

## Introduction

Slurry bubble columns are mass transfer and reaction devices in which gases enter at the bottom through a distributor as shown in Figure 1, then reacts on catalytic particles suspended in an inert liquid phase. The application of these reactors is of increasing importance in the petroleum and other industries. They have excellent mass and heat transfer characteristics for removal of the heat given off by exothermic reactions and the ability to replace catalyst easily. L-S Fan has given a thorough review of gas-liquid-solid fluidizations (*1*). An industrial review of gas-liquid-solid reactor development was given by Tarmy and Coulaloglou (*2*). Slurry bubble column reactors (SBCR) have recently become competitive with traditional fixed bed reactors for converting synthesis gas into liquid fuels (*3*). Major oil companies are gearing up to build SBCR to utilize natural gas located in remote areas of the world and to convert it to paraffin wax which will be upgraded to gasoline and Diesel fuels (*4, 5*). Stiegel (*6*) and Heydorn et al (*7*) published an excellent review of the Department of Energy (DOE) Research in Fischer-Tropsch technology and liquid phase methanol processes. They described the advantages of the slurry-phase reactor over the fixed bed reactor.

Early SBCR models were reviewed by Ramachandran and Chaudhari (*8*) and by Deckwer (*9*). They require hold-up correlations as an input and do not compute flow patterns. The most complete and useful of these models applied to the Fischer-Tropsch (F-T) conversion of synthesis gas in a SBCR is that of Prakash and Bendale (*10*). They sized commercial SBCR for DOE. They gave syngas conversion and production as a function of temperature, pressure and space velocity. Input parameters with considerable uncertainty that influenced production rates were the gas hold-up, the mass transfer coefficient and the dispersion coefficient. Krishna's group (*11*) extended such a model to compute product distribution using a product selectivity model. Air Products working with Dudukovic measured dispersion coefficients needed as an input into such model. The problem with this approach is that the dispersion coefficients are not constant. They are a function of the local hydrodynamics.

The multiphase computational fluid dynamics (CFD) approach does not require dispersion coefficients as an input. Hold-up and flow patterns are computed. The Dudukovic group is using the CFD approach to compute gas-liquid flow using Los Alamos CFDLIB code (*12*) with a viscosity as an input and are working on liquid-solid flow (*13*), but have no models or codes for SBCR. The L.S. Fan group built a unique high pressure bubble column and developed Particle Image Velocity (PIV) and other techniques useful for multiphase measurements. Y. T. Shah's group (*14*) and Rentech Inc. (http://www.rentechinc.com) are using the k-epsilon turbulence model in their

CFD approach. This model has too many unknown parameters when applied to multiphase flow. The kind of model used by Dudukovic group (*12*) computed the Reynolds stresses in agreement with measurements done in L.S. Fan's laboratory. In the Matonis et al (*15*) paper we show our capability to compute turbulence in a slurry bubble column in the churn-turbulent regime in agreement with our measurements.

Bukur's group (*16*) have used an autoclave reactor to obtain F-T kinetic data and improve catalysts.

Although considerable advances have been made both in understanding the basic aspects of slurry bubble column reactors (SBCR) and in developing rational design procedures, computational fluid dynamics (CFD) – assisted design methodology for reactor optimization is sparse.

Here, we have developed an algorithm for computing the optimum particle size for maximum production in a SBCR based on computational fluid dynamics (CFD) models. Reactor optimization due to heat exchanger configuration was also investigated. We have rearranged the heat exchangers in a SBCR and constructed a CFD model for a baffled reactor in the upper portion of the reactor. The baffle arrangement of the heat exchangers prevents the mixing of the catalyst from the upper stage allowing continued operation of the reactor with a high concentration in the upper stage. Thus an optimum catalyst concentration is maintained during the course of the production of liquid fuels.

## Slurry-Phase Reactor Vs Fixed-Bed Reactor

In a typical fixed-bed synthesis reactor (Figure 1a), the catalyst is packed into about 10,000 one-inch (2.5 cm) tubes, and the synthesis gas is sent through this array of tubes. The tubes are water-cooled, and much of the heat given off by the reactor is captured as process steam. However, heat transfer within the packed tubes is poor, and if a low $H_2$-to-CO synthesis gas were used, carbon would be deposited on the catalyst surfaces, resulting in severe catalyst deactivation and loss of operability.

In contrast to the fixed-bed reactor, the slurry-phase reactor (Figure 1b) is relatively simple. Basically, a finely divided catalyst is suspended in a liquid. For F-T synthesis, this liquid is a waxy portion of the product. The reactant gases are bubbled through the catalyst slurry and react on the surface of the catalyst to produce the desired product. The use of finely divided catalysts ensures high rates of reaction that result in a large amount of product per unit of reactor volume per unit of time. The presence of the liquid and the turbulence caused by the gas flow allow the effective removal of heat from the surface of the catalyst. Therefore, higher rates of reaction and greater conversion can be achieved.

In the slurry-phase reactor, the gas flow keeps the catalyst slurry well mixed. The heat given off by the reaction is recovered as steam through a heat exchanger embedded in the reactor volume. Moreover, the total heat exchanger tube area is only a small fraction of the tube area of a fixed-bed reactor for F-T synthesis. This excellent ability to remove the heat given off by the exothermic reaction prevents carbon from being deposited on catalyst surface, thereby reducing catalyst deactivation when low $H_2$-to-CO synthesis gas is used. These are the advantages of slurry-phase reactors.

The major uncertainty about using slurry-phase reactors (Figure 2) for F-T synthesis involves fluid flow and heat transfer characteristics in the reactor. DOE's slurry-phase reactor research effort will obtain hydrodynamics, heat transfer, and mass transfer data needed to properly design slurry-phase reactors. Optimization of slurry-phase reactor operation may require a number of process modifications, including changes in (1) reactor configuration to improve mixing, heat transfer, or vapor-liquid separation of the products; (2) catalyst particle size distribution or surface properties to reduce settling or to increase the concentration of catalyst within the reactor; and (3) the liquid composition in the reactor to allow for higher gas velocity or better heat transfer.

## Flow Regimes

In bubble column reactors the hydrodynamics, transport and mixing properties such as pressure drop, holdup of various phases, fluid-fluid interfacial areas, and interphase mass and heat transfer coefficients depend strongly on the prevailing flow regime. Shah et al (*17*) have reviewed the criteria that characterize the upward movement of the bubble swarms into three separate flow regimes. The regimes are identified depending on the increasing gas flow rate and column diameter. The main difference between these regimes is the bubble size distribution as summarized in Figure 3. These flow regimes are useful for the empirical correlations.

- Homogeneous flow or dispersed bubbles: This regime is characterized by almost uniformly sized bubbles with equal radial distribution. This regime occurs if the superficial gas velocity is less than 0.05m/s and the rise velocity of the bubbles lies between 0.18 and 0.30 m/s. The theory of bubbly flow was developed by several investigators. The theory fails if significant mass transfer occurs in the column. Using stagnant bubble clouds in a flowing liquid, the bubble flow regime can be realized up to gas holdup of 66%, whereas in the usual arrangement with almost stagnant liquid the bubbly flow theory fails of the gas holdup is larger than about 15%.

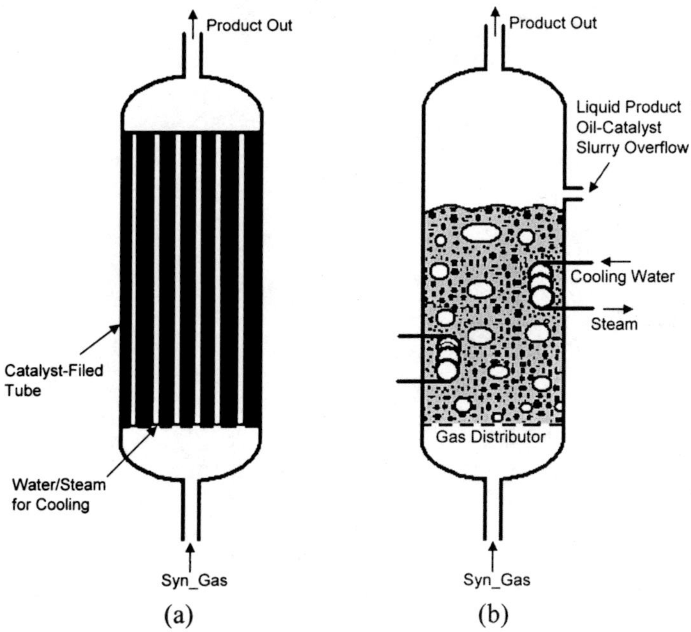


*Figure 1. Reactor used for Fischer-Tropsch synthesis: (a)fixed bed reactor, (b) slurry phase reactor*

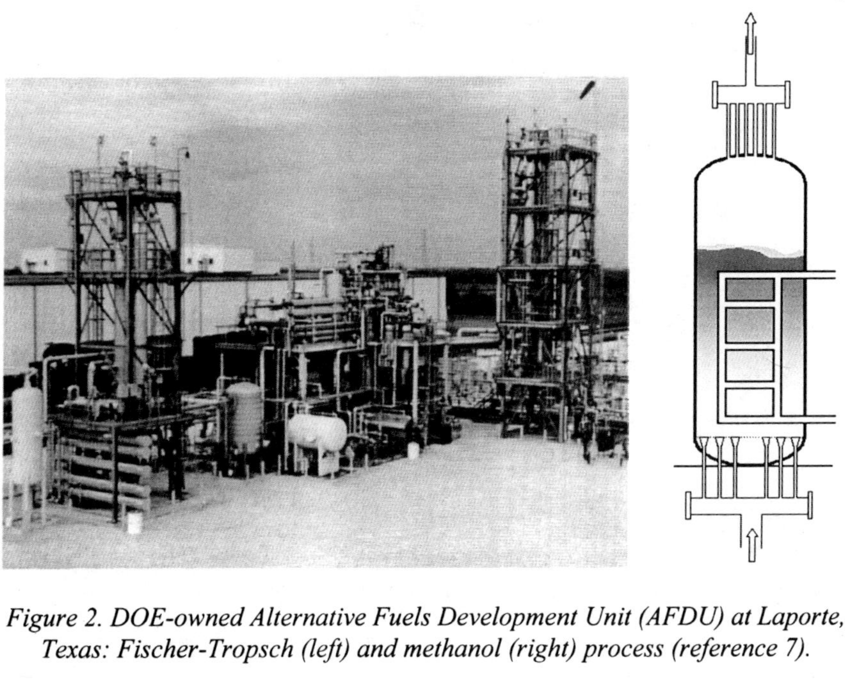

*Figure 2. DOE-owned Alternative Fuels Development Unit (AFDU) at Laporte, Texas: Fischer-Tropsch (left) and methanol (right) process (reference 7).*

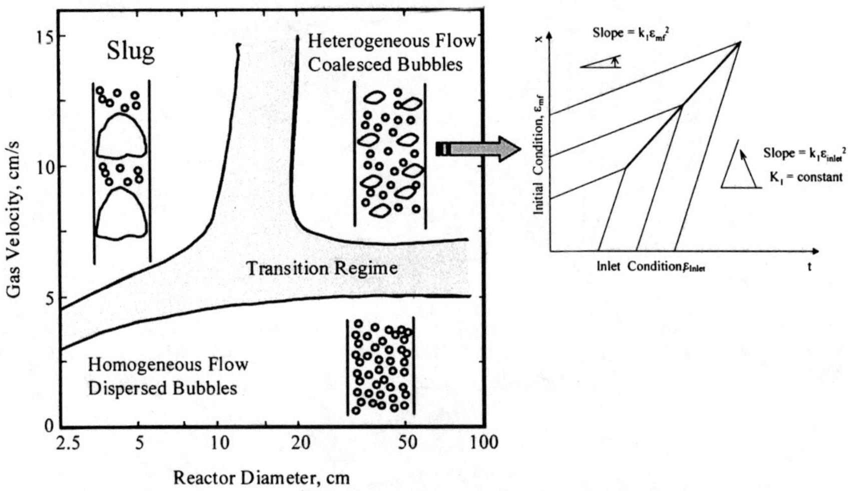


*Figure 3. Flow regimes and bubble formation in a bubble column reactor and bubble formation (references 17, 18).*

- Heterogeneous (churn turbulent) flow or coalesced bubbles: At higher gas velocities the homogeneous gas-in- liquid dispersion cannot be maintained and an unsteady flow pattern with channeling occurs. This heterogeneous flow regime is characterized by large bubbles moving with high rise velocities in the presence of small bubbles. The large bubbles take the form of spherical caps with a very mobile and flexible interface. These large bubbles can grow up to a diameter of about 0.15 m.

- Slug flow: In small diameter columns, at high gas flow rates, large bubbles are stabilized by the column wall leading to the formation of bubbles slugs. Bubble slugs can be observed in columns of diameters up to 0.15 m.

The type of sparger used, physico-chemical properties of liquid, the liquid velocity can also affect the transition between the flow regimes. The boundaries between the different flow regimes shown in Figure 3 are only approximate.

The bubble coalescence criterion (Figure 3) using the propagation velocity was developed by Gidaspow (*18*). The void propagation equation in one dimension was derived from the continuity equations. The propagation velocity (C) in coalesced bubble regime becomes as follows:

$$C = \frac{\Delta \rho g}{150 \mu_{mixture}} \varepsilon^2 d_{bubble}^2 = (\nabla \rho g)^{\frac{1}{3}} \frac{(6 \sigma D_{orifice})^{\frac{2}{3}}}{150 \mu_{mixture}} \varepsilon^2 \qquad (1)$$

In dispersed bubble regime

$$C = \frac{\Delta \rho g d_{bubble}^2}{18 \mu} \qquad (2)$$

## Traditional Approach using Hold-up and Correlations

Prakash and Bendale *(10)* have undertaken an extensive literature search on gas holdup and dispersion correlations in slurry bubble column reactors. They tested various correlations for the estimation of gas holdup against experimental data and found that the average absolute relative error varied from 12 to 165%. These large errors are primarily due to the fact that literature correlations for gas holdup are based on constant gas velocity along the column height. In reality, gas velocity decreases significantly along the reactor height due to decrease in moles of products for methanol synthesis. The computational fluid dynamic approach describes in this chapter include the effect of decreasing gas velocity on gas holdup and other hydrodynamic parameters.

## Computational Fluid Dynamics Approach

The hydrodynamic approach to multiphase flow systems is based on the principles of mass conservation, momentum balance and energy conservation for each phase (*18*). The kinetic theory based multiphase CFD model that describes the hydrodynamics of SBCR was developed (*19, 20*). This model includes the complete granular temperature balance based on the kinetic theory of granular flow. Stuart Savage and others (*21, 22*) showed that the dense-phase kinetic theory, as presented by Chapman and Cowling (*23*), can be applied to granular flow of particles. Gidaspow (*18, 24*) has reviewed this theory. This model treated the particle phase as another fluid with its own temperature, called the granular temperature, its own pressure due to particle collision and its own viscosity. The granular temperature, which is like the thermal temperature in kinetic theory of gases, measures the random oscillations of particles. The particle pressure and the particle viscosity are a function of the granular temperature, which varies with time and position in a fluidized bed. The CFD model is summarized in Table 1. Gamwo et al (20) have provided detailed nomenclature of the model. In the approach of Ahmadi and Ma (*25*) and Cao and Ahmadi (*26*), a fluctuating kinetic energy balance (Eq. T17) is written for each phase. For the gas-solid system they find the reasonable result that the fluctuating velocity of particles is the same as that of the fluid. This assumption was made in this study. Hence only the equation for the granular temperature of the particles is needed. Constitutive equations for solids pressure and viscosity, and drag coefficients were described in Gidaspow's book (*18*) and in Gamwo, et al (*20*).

The reactions and solubilities were given in the Viking report (*10*). The detailed reactions and mass transfer for methanol production in the LaPorte pilot plant were described by Gamwo, et al (*20*).

### Reactions:

A review of the literature showed that the following chemical reactions are accepted for production of methanol from syn-gas. The reactions included are methanol production from hydrogen and CO, water gas shift reaction, and the production of methanol from $CO_2$-hydrogenation.

$$
\begin{array}{lllllll}
CO & + & 2H_2 & \leftrightarrow & CH_3OH & & \\
CO_2 & + & H_2 & \leftrightarrow & CO & + & H_2O \\
CO_2 & + & 3H_2 & \leftrightarrow & CH_3OH & + & H_2O
\end{array} \tag{3}
$$

Graaf et al (*27,28*) developed the reaction rate for methanol synthesis in gas-catalyst phases and extended it to three phase methanol synthesis using gas-liquid solubilities in thermodynamical equilibrium described by Henry's law. The rates for the three reactions ($IX = 3$) in liquid phase are given by equations T12 and T13. Where, $\alpha_i^{jx}$ represents the stoichiometric coefficient of *jx*th species in the *i*th reaction of liquid phase and $M^{jx}$ represents the molecular weight of *jx*th species. $r''_{IX}$ (mol/Kgcat·sec) is the rate of reaction on catalyst surface for three reactions.

$$
r''_{CH3OH,A3} = \frac{k^*_{ps,A3} k^*_{CO} \left(C_{CO} C_{H2}^{3/2} - C_{CH3OH} / \left(C_{H2}^{1/2} K_{C1}\right)\right)}{\left(1 + k^*_{CO} C_{CO} + k^*_{CO2} C_{CO2}\right)\left(C_{H2}^{1/2} + \left(k^*_{H2O} / k^{*1/2}_{H2}\right) C_{H2O}\right)} \tag{4}
$$

$$
r''_{H2O,B2} = \frac{k^*_{ps,B2} k^*_{CO2} \left(C_{CO2} C_{H2} - C_{H2O} C_{CO} / K_{C2}\right)}{\left(1 + k^*_{CO} C_{CO} + k^*_{CO2} C_{CO2}\right)\left(C_{H2}^{1/2} + \left(k^*_{H2O} / k^{*1/2}_{H2}\right) C_{H2O}\right)} = r''_{co,B2} \tag{5}
$$

$$
r''_{CH3OH,C3} = \frac{k^*_{ps,C3} k^*_{CO2} \left(C_{CO2} C_{H2}^{3/2} - C_{CH3OH} C_{H2O} / \left(C_{H2}^{2/3} K_{C3}\right)\right)}{\left(1 + k^*_{CO} C_{CO} + k^*_{CO2} C_{CO2}\right)\left(C_{H2}^{1/2} + \left(k^*_{H2O} / k^{*1/2}_{H2}\right) C_{H2O}\right)} = r''_{H2O,C3} \tag{6}
$$

where, chemical equilibrium constants are

$$
K_{C1} = 1.72 \times 10^{-16} e^{\left(\frac{126011}{RT}\right)}, \quad K_{C2} = 5.81 \times 10 e^{\left(\frac{-33760}{RT}\right)}, \quad K_{C3} = K_{C1} \times K_{C2} \tag{7}
$$

Reaction rate constants are

$$
\begin{gathered}
k^*_{ps,A3} = 1.66 \times 10^5 e^{\left(\frac{-93925}{RT}\right)}, \quad k^*_{ps,B2} = 7.21 \times 10^{17} e^{\left(\frac{-215130}{RT}\right)}, \quad k^*_{ps,C3} = 8.52 \times 10^{-1} e^{\left(\frac{-43425}{RT}\right)}, \\
k^*_{CO} = 9.01 \times 10^{-12} e^{\left(\frac{92138}{RT}\right)}, \quad k^*_{CO2} = 3.15 \times 10^{-5} e^{\left(\frac{34053}{RT}\right)}, \\
k^*_{H2O} / k^*_{H2^{1/2}} = 2.71 \times 10^{-12} e^{\left(\frac{103030}{RT}\right)}
\end{gathered} \tag{8}
$$

$C_{jx}$ (mol/m$^3$) is the bulk concentration of the *jx*th species in the liquid phase and R is the gas constant of 8.314 (J/mol-K).

**Table.1 Hydrodynamic kinetic theory model for SBCR**

*Conservation of Mass for Each Phase*

Gas phase:

$$\frac{\partial}{\partial t}(\varepsilon_g \rho_g) + \nabla \cdot (\varepsilon_g \rho_g v_g) = \dot{m}_g \tag{T1}$$

Liquid and solid phase: ($k = l$, s)

$$\frac{\partial}{\partial t}(\varepsilon_k \rho_k) + \nabla \cdot (\varepsilon_k \rho_k v_k) = \dot{m}_k \tag{T2}$$

where $\dot{m}_l = \varepsilon_l \sum_j^N M^j R_j$, $\dot{m}_g = -\dot{m}_l$, $\dot{m}_s = 0$, $R_j$ is the rate of mass transfer between phases.

*Momentum Equations*

Gas phase:

$$\frac{\partial}{\partial t}(\varepsilon_g \rho_g v_g) + \nabla \cdot (\varepsilon_g \rho_g v_g v_g) = \varepsilon_g \rho_g F_g + \sum_{m=l,s} \beta_{gm}(v_m - v_g) + \nabla \cdot [\tau_g] + \dot{m}_g v_g \tag{T3}$$

Liquid and solid phase: ($k = l$, s)

$$\frac{\partial}{\partial t}(\varepsilon_k \rho_k v_k) + \nabla \cdot (\varepsilon_k \rho_k v_k v_k) = \varepsilon_k \rho_k F_k + \sum_{m=g,l,s} \beta_{km}(v_m - v_k) + \nabla \cdot [\tau_k] + \dot{m}_k v_k \tag{T4}$$

*Energy Equations*

Gas phase:

$$\frac{\partial}{\partial t}(\rho_g \varepsilon_g H_g) + \nabla \cdot (\rho_g \varepsilon_g H_g v_g) = (\frac{\partial P_g}{\partial t} + v_g \cdot \nabla P_g) + \nabla \cdot (k_g \nabla T_g) + \sum_i r_{ig} \Delta H_{ig} + \sum_{m=\ell,s} \{h_{vm}(T_m - T_g) + \beta_{gm}(v_m - v_g)^2 - \dot{m}_m H_g\} \tag{T5}$$

Liquid and solid phase: ($k = l$, s)

$$\frac{\partial}{\partial t}(\varepsilon_k \rho_k H_k) + \nabla \cdot (\varepsilon_k \rho_k H_k v_k) = \nabla \cdot (k_k \nabla T_k) + \sum_i r_{ik} \Delta H_{ik} + h_{vk}(T_g - T_k) + \sum_{m=g,l,s} \beta_{km}(v_m - v_k)^2 + \dot{m}_k H_g \tag{T6}$$

**Table 1. *Continued.***

*Fluctuating Energy Equation for the Particles: (k = s)*

$$\frac{3}{2}\left[\frac{\partial}{\partial t}(\varepsilon_k\rho_k\theta_k)+\nabla\cdot(\varepsilon_k\rho_k\theta_k v_k)\right]=\nabla\cdot(\kappa_k\nabla\theta_k)-\gamma_k+\Phi_k \qquad \text{(T7)}$$

*Constitutive Equations*

Shear stresses:

$$[\tau_g]=\left\{-P_g-\frac{2}{3}\mu_g\varepsilon_g\nabla\cdot v_g\right\}[I]+\mu_g\varepsilon_g\left[\nabla v_g+(\nabla v_g)^T\right]$$

$$[\tau_k]=\left\{-P_k+(\xi_k-\frac{2}{3}\mu_k)\nabla\cdot v_k\right\}[I]+\mu_k\left[\nabla v_k+(\nabla v_k)^T\right]\ \mathrm{k}=l,s \qquad \text{(T8)}$$

Gas-solid heat transfer: (Gunn's model)

$$\mathrm{Nu}_k=\left\{(2+5\varepsilon_k^2)\left(1+0.7\,\mathrm{Re}_k^{0.2}\,\mathrm{Pr}^{\frac{1}{3}}\right)+\left(\frac{2}{15}+1.2\varepsilon_k^2\right)\mathrm{Re}_k^{0.7}\,\mathrm{Pr}^{\frac{1}{3}}\right\}\mathrm{Sp}_k$$

$$\mathrm{Nu}_k=\frac{h_{vk}d_k}{k_g^0},\qquad \mathrm{Pr}=\frac{C_{P_g}\mu_g}{k_g^0},\qquad \mathrm{Sp}_k=\frac{6\varepsilon_k}{d_k} \qquad \text{(T9)}$$

*Reactions, Kinetics and Mass Transfer*

Gas phase species balances:

$$\frac{\partial}{\partial t}(\varepsilon_g\rho_g y_g^j)+\nabla\cdot(\varepsilon_g\rho_g y_g^j v_g)=-\varepsilon_l M^j R_j \qquad \text{(T10)}$$

Liquid phase species balances:

$$\frac{\partial}{\partial t}(\varepsilon_k\rho_k y_k^j)+\nabla\cdot(\varepsilon_k\rho_k y_k^j v_k)=\varepsilon_l M^j R_j \qquad \text{(T11)}$$

Reactions of *jx*th species of liquid phase :

$$r_l^{jx}=\sum_{i=1}^{IX}\alpha_i^{jx}M^{jx}r_i^{'} \qquad \text{(T12)}$$

Where $r_{IX}^{'}=\frac{\varepsilon_s\rho_s}{1.0\times10^3}\cdot r_{IX}^{''}(\frac{mol}{cm^3\cdot\sec})$ (T13)

Mass transfer between gas and liquid:

$$\dot{m}_k=\sum_{jx=1}^{n}\dot{m}_k^{jx}M^{jx} \qquad \text{(T14)}$$

where $\dot{m}_l^{jx}=\varepsilon_l k_l a\left(C_{jx}^{g-l}-C_{jx}\right)\times\frac{1}{10^6}(\frac{mol}{cm^3\cdot\sec})$ (T15)

$$\dot{m}_l=\sum_{jx=1}^{n}\dot{m}_l^{jx}M^{jx}=-\dot{m}_g,\qquad \dot{m}_s=0 \qquad \text{(T16)}$$

**Mass Transfer:**

The rate of mass transfer was given by equation T14. Mass transfer between gas phase and liquid phase (Eq. T16) was only considered in this study. Mass transfer rate (Eq. T15) between gas phase and liquid phase can be expressed in terms of the volumetric mass transfer coefficient and the concentration difference between gas-liquid interface phase and liquid phase. Where, $k_l a$ (sec$^{-1}$) is the volumetric mass transfer coefficient and $C^{g\text{-}l}_{jx}$ is the concentration of the *jx*th species at gas-liquid interface phase which can be defined by Henry's law. The fugacity of a very dilute species in a liquid phase is linearly proportional to its mole fraction at low mole fractions.

$$f_{jx} = H_{jx} C^{g-l}_{jx} \,(\text{mol/m}^3) \tag{9}$$

In gas-liquid phase equilibrium, the fugacity of liquid phase can be the fugacity of gas phase defined by partial pressure of species in the gas phase.

$$f_{jx} = y'_{jx} P \,(\text{bar}) \tag{10}$$

where, $f_{jx}$ (bar) is the Fugacity of the *jx*th species, $y'_{jx}$ is the mole fraction of *jx*th species and $H_{jx}$ (bar·m$^3$/mol) is the Henry's constant of the *jx*th species used by Graaf et al (*27, 28*).

$$H_{CO} = 0.175 e^{\left(\frac{638}{RT}\right)},\ H_{CO2} = 0.402 e^{\left(\frac{-6947}{RT}\right)},\ H_{H2} = 0.0782 e^{\left(\frac{4875}{RT}\right)}$$

$$H_{H2O} = 0.330 e^{\left(\frac{-8633}{RT}\right)},\ H_{CH3OH} = 1.49 e^{\left(\frac{-17235}{RT}\right)} \tag{11}$$

## Recent Experimental Hydrodynamic Measurements

Tartan and Gidaspow (*29*) have recently developed an experimental kinetic theory based particle image velocity technique for measuring particle and Reynolds stresses in gas-solid risers. They have shown that for gas-solid flow that are two types of turbulence in the risers: random oscillations of the individual particles and oscillations of clusters measured by the Reynolds stresses of the particles. Earlier Mudde et al (*30*) have obtained similar measurements for bubble columns. Pan et al (*12*) have compared Mudde et al (*30*) experiments to simulations using the Los Alamos CFD code. Figure 4 shows typical Reynolds stress computations to the experiments.

Figure 5 shows an example of the simulations based on the Bahary's (*31*) experimental data in the IIT slurry bubble column. The instantaneous particle velocity fluctuates upward and downward as visually confirmed in the experiment. There is more vigorous fluctuation of particles in the center region. This flow structure due to fluctuating particle velocity causes the particle concentration to be uniform throughout the bed. Hence it gives good mixing in the slurry bubble column. The flow patterns agree with the literatures (*32, 18*). Two distinct bubbles seen in the middle of the bed were captured, as the relative higher velocity in the simulations. This CFD simulation was employed to compute the Reynolds stresses and the granular-like temperature to IIT experimental data. Matonis et al (*15*) and Jung (*33*) showed the reasonable computational results, in agreement with experimental results. The granular-like temperature was calculated from time-average Reynolds normal stresses measured by the particle image velocity technique.

$$\theta_{granular-like} = \frac{1}{3}\overline{v_x^{'}v_y^{'}} + \frac{1}{3}\overline{v_y^{'}v_y^{'}} + \frac{1}{3}\overline{v_z^{'}v_z^{'}} \tag{12}$$

## Optimization using Computational Fluid Dynamics

The more advanced model of Wu and Gidapow (*19*) is used to explore novel reactor designs: optimum catalyst size and reactor configuration. The model included the effect of the mass transfer coefficient between the liquid phase and the gas phase and the water-gas shift reaction. With reaction, this model was used to predict slurry height, gas hold-up and the rate of methanol production of the Air Products/DOE LaPorte slurry bubble column reactor.

An issue of interest to the energy industries throughout the world is the size of catalyst they should make for SBCR (*5, 34*). Catalyst particles as a powder form used in most fluidized bed process are small enough for external mass transfer and internal diffusion resistance to be negligible. The size of the catalyst is typically in the range of 20μm to 120μm (*35, 36, and 37*). Small particle sizes are needed to have good effectiveness for reaction. However, small particles are entrained in the product gas stream and are known to cause liquid product filtration problems. Small particles also cause the formation of the clusters, which give large effective particle sizes and hence poor mass transfer. Here we have applied the mathematical model for gas-liquid-solid flows to determine the optimum particle size, which is the size that has the maximum granular temperature, similar to the experiments for gas-solid systems done at EXXON (*38*). For this particle size, the heat and the mass transfer coefficients have the

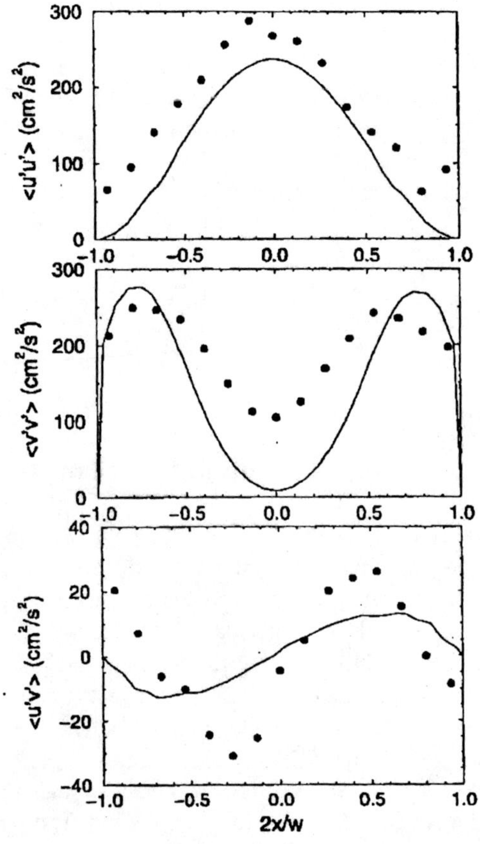


*Figure 4. Time-averaged profiles of turbulence intensities, <u'u'> and <v'v'>, and Reynolds shear stress, <u'v'>, for the middle section of 15 cm bubble column at $U_{sup}$ = 1 cm/s. (Reproduced with permission from reference 12. Copyright 2000.)*

highest values. In this model the mass transfer coefficient is an input. We have related the mass transfer coefficient to the fluctuating velocities (granular temperature) computed by the hydrodynamic model.

SBCR, such as those used to produce methanol and other liquids from synthesis gas, normally have a high catalyst concentration at the bottom of the reactor, when there is no liquid recirculation. M. Chang and C.A. Coulaloglu (*39*) of EXXON describe the disadvantages of such a catalyst distribution and suggest an improvement. Wu and Gidaspow (*19*) developed a CFD model for such process and compared the CFD results to the La Porte pilot plant methanol production runs. The CFD results show the unfavorable high solids concentration at the bottom of the reactor. Here we have rearranged the heat exchangers in the La Porte unit and constructed a CFD model for a baffled reactor that has a higher concentration of the catalyst in the upper portion of the reactor (*40*).

## Optimum Catalyst Size

Mass transfer coefficient was estimated from a relationship between the fundamental equations of the boundary layers and the turbulent kinetic energy of particles (granular temperature) computed by the hydrodynamic model with no reaction (*20*). We have varied the particle size from 20 μm to 100 μm and discovered a maximum in the granular temperature (Figure 6). For the particles over this range, the mass transfer coefficient has the highest values (*20*). The maximum granular-like temperature is at 60μm between 50 μm and 75μm, with a solid loading of about 10%. It agrees well with the experimental results of Cody's study (*38*) for gas-solid bubbling bed. They showed that Geldart A glass spheres exhibit an order magnitude higher granular temperature than neighboring Geldart B glass spheres based on the experiments in a gas-solid bubbling bed.

Gamwo et al (*20*) also estimated the effective particle diameter from the coagulation theory (*41*). The volumetric mass transfer coefficient based on this effective particle diameter shown in Figure 7 has a maximum value near the particle diameter of 50 μm. It may be larger than the estimated value due to non-spherical particle effects or electrical forces. As pointed out previously, small particles cause the formation of the clusters, which give large effective particle sizes and hence poor mass transfer. Large particles, Geldart B particles, have very low mass transfer coefficients and hence poor production rate. Our results show a good agreement with that of a slurry bubble column reactor in the Viking Systems International Report given by Prakash and Bendale (*10*).

A search for an optimum catalyst size was carried out for methanol synthesis in the DOE Laporte SBCR using the complete kinetic theory model of granular flow. Figure 8 show an example for the optimization of slurry-phase reactor operation based on LaPorte SBCR. The mole fraction of methanol and

(a)

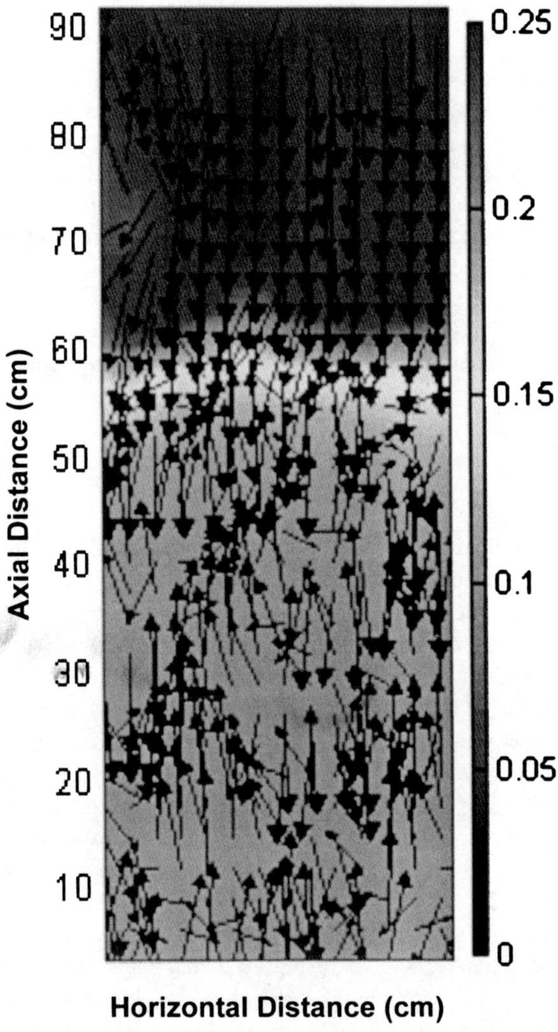


*Figure 5. Three-dimensionsal simulation for 800 μm particles with a viscosity input model in the IIT slurry bubble column with a 5 cm width: (a) instantaneous particle velocity vector plots and solid volume fractions (color bar) at 25 s, and (b) time-averaged particle velocity at the center (c) near the wall from 15 to 36 s. ($V_L$=2.02cm/s, $V_G$=3.37cm/s)*

*(See page 1 of color inserts.)*

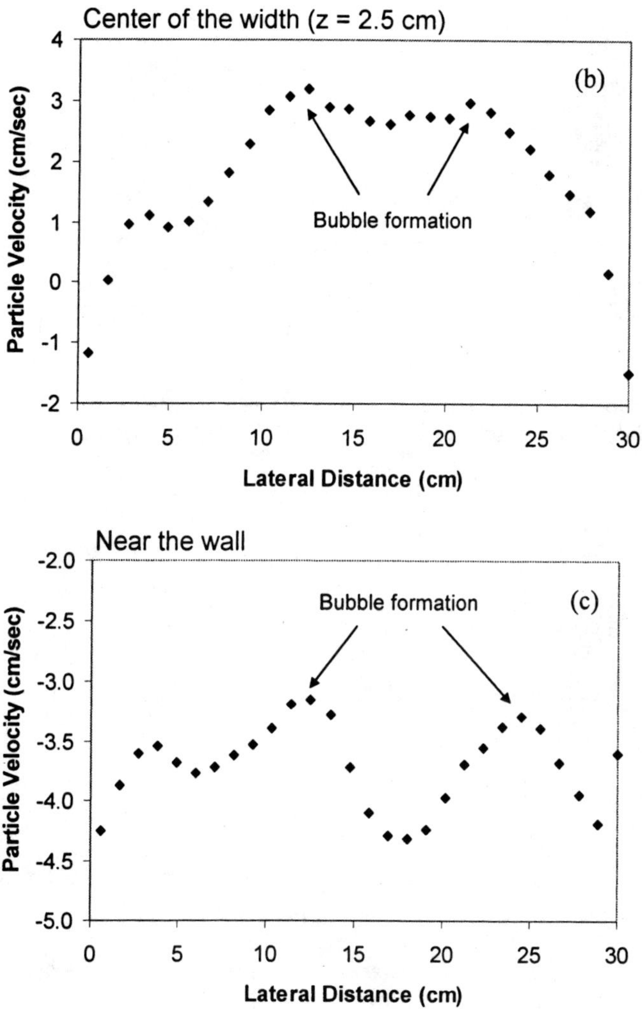


*Figure 5. Continued.*

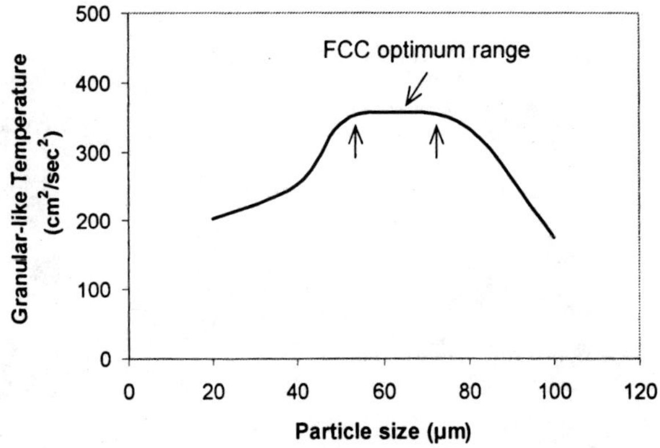


*Figure 6. Optimum particle size for mixing (maximum granular-like temperature) in the IIT slurry bubble column. ($V_L$=2.02cm/sec, $V_G$=3.37cm/sec)*

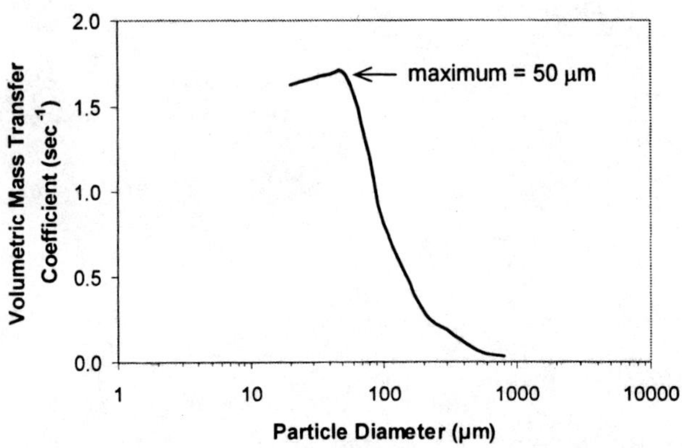


*Figure 7. Volumetric mass transfer coefficient based on effective particle diameter using coagulation theory.*

(a)

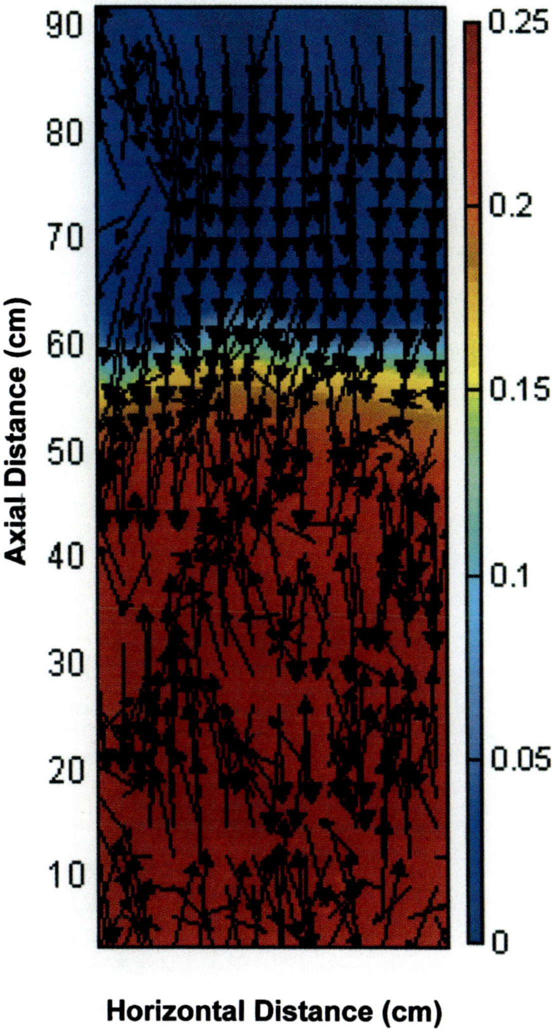


*Figure 17.5. Three-dimensionsal simulation for 800 μm particles with a viscosity input model in the IIT slurry bubble column with a 5 cm width: (a) instantaneous particle velocity vector plots and solid volume fractions (color bar) at 25 s, and (b) time-averaged particle velocity at the center (c) near the wall from 15 to 36 s. ($V_L$=2.02cm/s, $V_G$=3.37cm/s)*

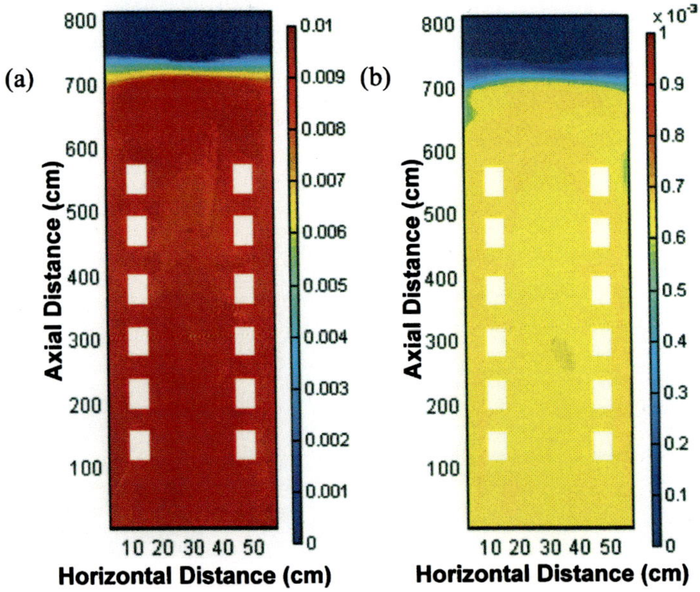


*Figure 17.9. Computed mole fraction of (a) methanol and (b) water in the liquid phase at 30 s in the simulation of LaPorte's methanol synthesis using the kinetic theory.*

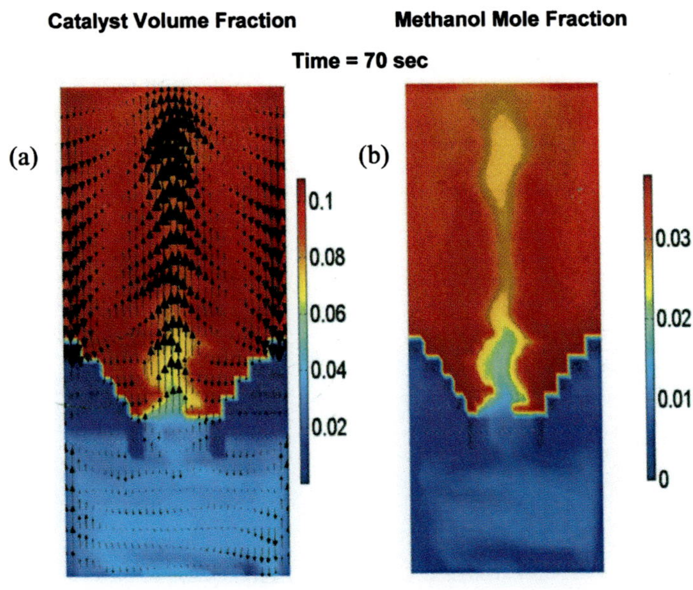


*Figure 17.13. Preferred reactor operation with a high catalyst concentration in the upper portion of the reactor (Reproduced with permission from reference 40. Copyright 2003.)*

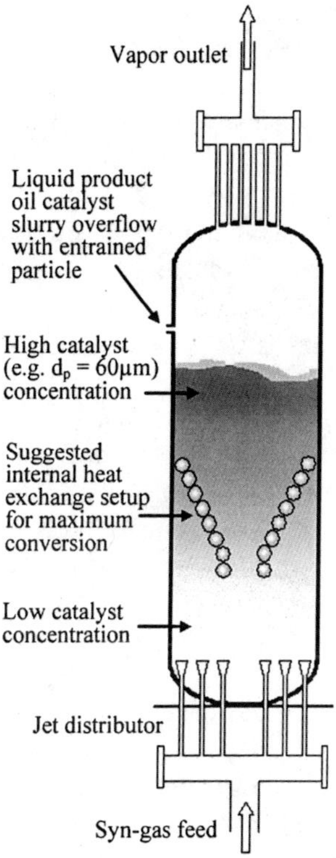


Initial conditions and system properties

| | |
|---|---|
| Diamter of reactor (cm) | 61 |
| Height of reactor (cm) | 853 |
| Temperature (°C) | 250.3 |
| Pressure (psig) | 753.2 |
| Inlet gas velocity (cm/s) | 15.24 |
| Diameter of catalyst (μm) | 50 |
| Density of catalyst (Kg/m$^3$) | 3011 |
| Liquid | Wax |
| Density of liquid (Kg/m$^3$) | 851.5 |
| Initial slurry height (cm) | 183 |
| Initial liquid height (cm) | 427 |

Syn-gas composition (CO-rich)

| | CO | $CO_2$ | $H_2$ | $CH_3OH$ | $H_2O$ | $N_2$ | Wax |
|---|---|---|---|---|---|---|---|
| Mol % | 51.00 | 13.00 | 35.00 | 0.0 | 0.0 | 1.00 | 0.0 |
| Wt % | 68.07 | 27.26 | 3.34 | 0.0 | 0.0 | 1.33 | 0.0 |

*Figure 8. The SBCR with preferred heat exchanger arrangement and catalys particle size, and system properties for simulations.*

water in the liquid phases for the LaPorte SBCR are shown in Figure 9. The product water concentration is small, since we assumed the inlet synthesis gas was dry. With the water gas shift reaction, the ratio of $H_2$ to CO is 0.5 due to chemical equilibrium in the liquid phase of the simulation.

Figure 10 shows a comparison of the computed granular temperature and solid viscosity to experimental data for 45 μm Air Products methanol catalyst particles (*42*). The average computed granular temperature for the LaPorte SBCR was around 30 $cm^2/sec$ (*20*). It was close to that measured in the slurry bubble column at IIT. The average computed catalyst viscosity was close to 1 cp. It was much higher around the heat exchanger tubes due to the higher granular temperature at this location. The mixing in the reactor was very good.

Figure 11 shows the effect of the volumetric mass transfer coefficient estimated by the simulation in SBCR without liquid circulation. Methanol production increases 6.5 mol/kgcat·hr for the volumetric mass transfer coefficient of 0.75 and then the methanol production is no longer limited by volumetric mass transfer coefficient. This trend matched with that given in the Viking (*10*) report. The estimated volumetric mass transfer coefficient has a good agreement with experimental values shown in literatures (*10, 43, and 44*).

Figure 12 shows the methanol production for five different catalyst sizes. For 75 microns, the production has increased, then decreases substantially for 100 microns. Hence the optimum particle size is about 70 microns. This agrees with the previous simulations of about 60 microns for the case of no reaction.

## Reactor Optimization due to Heat Exchanger Configuration

In the present La Porte reactor (Figure 2) the basic flow pattern is an upflow at the center and downflow at the walls. This type of a flow pattern mixes the product with the reactants at the bottom of the reactor and causes poorer production of products.

The heat exchangers in the Air Products/DOE La Porte SBCR were rearranged as shown in Figure 8 (*40*). At start up the solids concentration at the bottom of the reactor was assumed to be high, close to the minimum fluidization concentration. When the gas flow was turned on the initially high solids concentration flowed into the upper portion of the reactor. The continued flow of the reactant gas prevented the return of the catalyst into the lower stage. A CFD video of the simulation shows these phenomena (Figure 13). In this new arrangement the conversion to products (8.7 mol/kgcat·hr) is higher than in the La Porte unit, because there is more catalyst in the region of decreased reactant concentration. The baffled arrangement of the heat exchangers prevents the mixing of the catalyst from the upper stage, allowing continued operation of the reactor with a high concentration in the upper stage. Thus an optimum catalyst concentration is maintained during the course of the production of the liquid fuels.

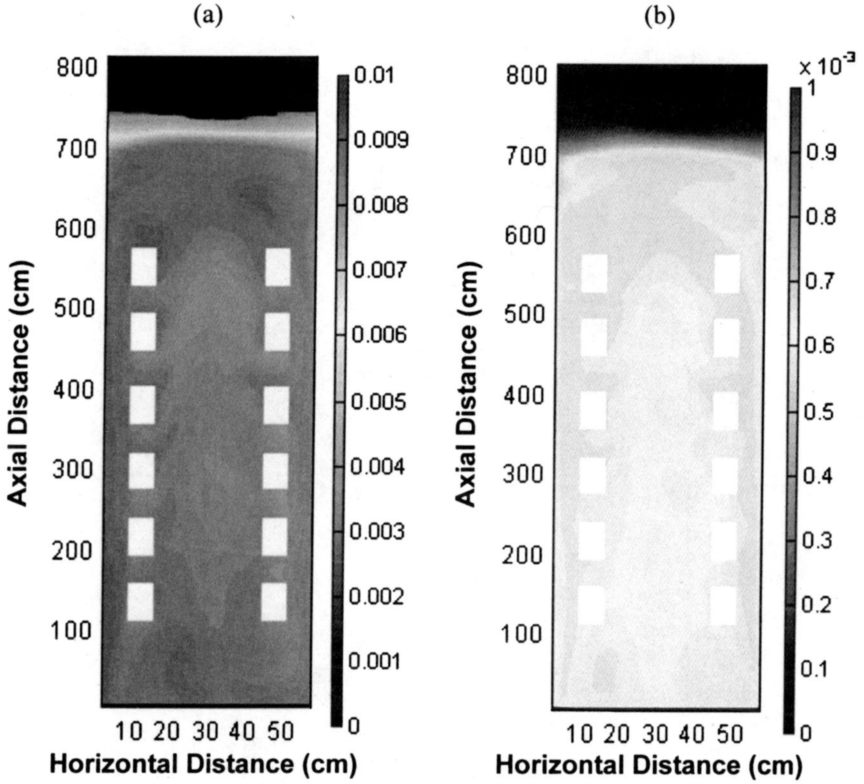


*Figure 9. Computed mole fraction of (a) methanol and (b) water in the liquid phase at 30 s in the simulation of LaPorte's methanol synthesis using the kinetic theory.* *(See page 2 of color inserts.)*

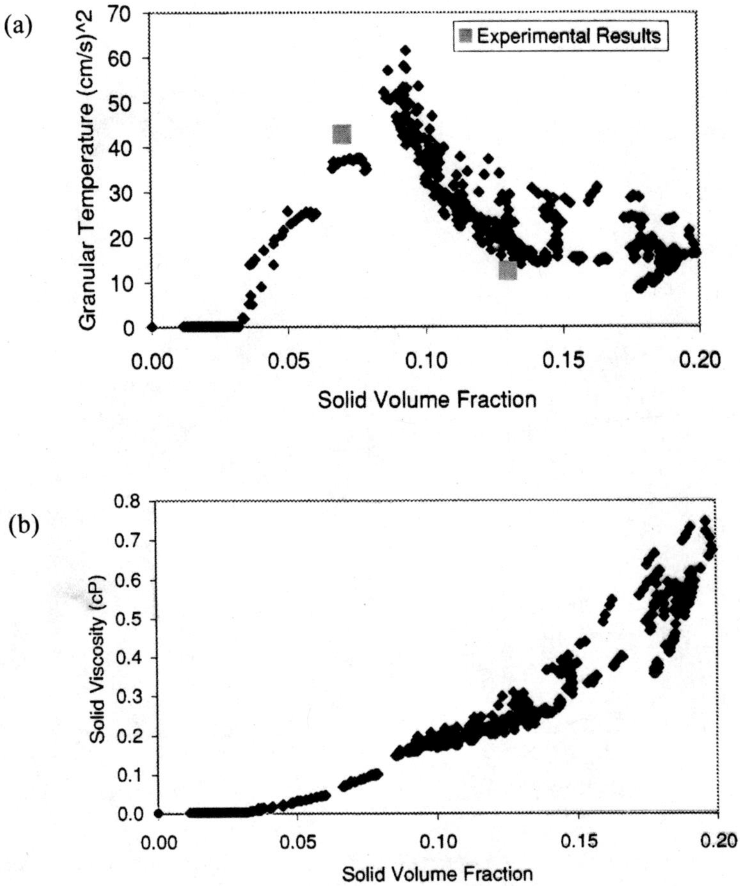


*Figure 10. Calculated (a) granular temperature and (b) solid viscosity compared to experimental data for 45 μm Air Products methanol catalyst particles measured in IIT's slurry bubble column. (Reference 41)*

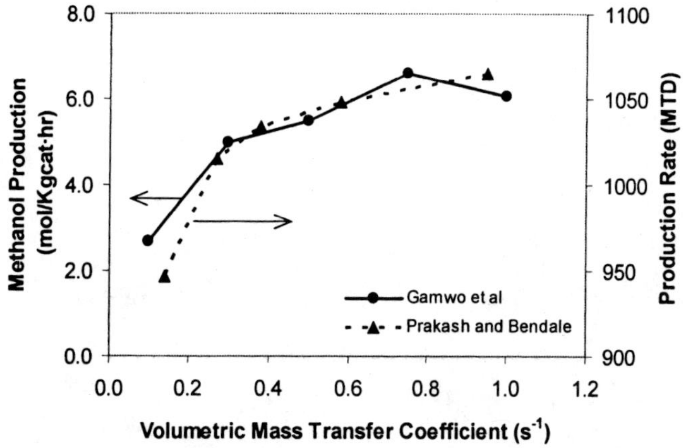


*Figure 11. Comparison of methanol production in the simulation of LaPorte's SBCR for various volumetric mass transfer coefficients. (Data are from references 10, 20)*

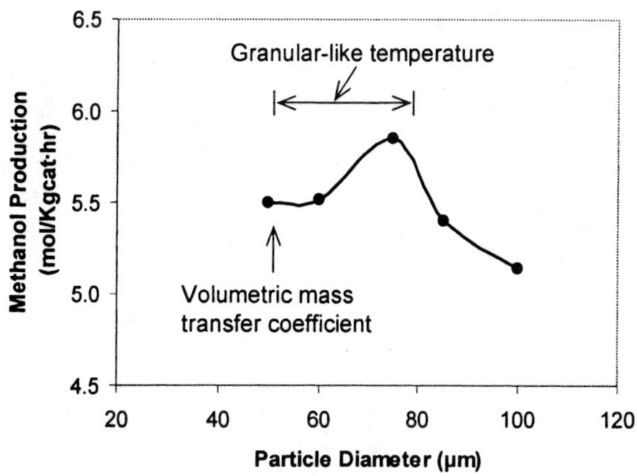


*Figure 12. Methanol production for five different catalyst sizes in the simulation of LaPorte's SBCR.*

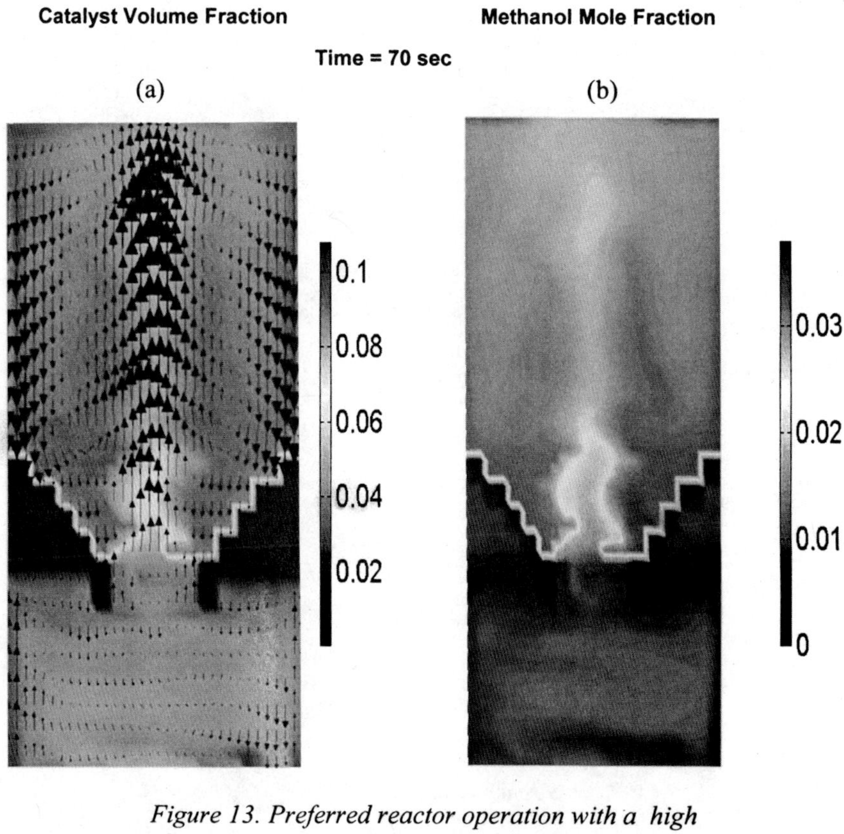


*Figure 13. Preferred reactor operation with a high catalyst concentration in the upper portion of the reactor (Reproduced with permission from reference 40. Copyright 2003.) (See page 2 of color inserts.)*

# Conclusions and Future Directions

## Conclusions

Slurry bubble column reactor for methanol and other hydrocarbons productions from synthesis gas is an issue of interest to the energy industries throughout the world. Computational fluid dynamics (CFD) is a recently developed tool which can help in the scale up. We have developed an algorithm for computing the optimum process of fluidized bed reactors. The mathematical technique can be applied to gas-solid, liquid-solid, and gas-liquid-solid fluidized bed reactors, as well as the LaPorte slurry bubble column reactor. Our computations for the optimum particle size show that there is a factor of about two differences between 20 and 60 μm size with maximum granular-like temperature (turbulent kinetic energy) near the 60 μm size particles.

The kinetic theory model[1] was extended to include the effect of the mass transfer coefficient between the liquid and the gas and the water gas shift reaction in the slurry bubble column reactor. The computed granular temperature was around 30 $cm^2/sec$ and the computed catalyst viscosity was closed to 1.0 cp. The volumetric mass transfer coefficient estimated by the simulation has a good agreement with experimental values shown in the literature. The optimum particle size was determined for maximum methanol production in a SBCR. The size was about 60 - 70 microns, found for maximum granular temperature. This particle size is similar to FCC particle used in petroleum refining.

Reactor optimization due to heat exchanger configuration was also investigated. The concentration of the catalyst in the upper stage was higher than in the lower stage due to the fact that the catalyst is not allowed to return to the lower stage. The methanol concentration was higher in the upper stage due to the continued production and flow from the lower stage.

## Future Directions

More than a quarter century ago NASA sponsored work has demonstrated that it is possible to compute single phase turbulent flow, the logarithmic velocity profile and the Reynolds stresses, by direct numerical solution of the Navier-Stokes equations for reasonably low Reynolds numbers, of the order of 10,000 (*45*). Recent research (*46*) has shown that it is similarly possible to compute the turbulent properties of dense gas-solid flow. For the dense riser flow, the computed energy spectrum captured the observed gravity wave and the Kolmogorov -5/3 law at high frequencies. The CFD simulations compared reasonably well with the measured core-annular flow experiments at high solid fluxes. The computed granular temperatures, solids pressures, FCC viscosities and frequencies of particle oscillations were close to the experiments reported in the literature. Here we show the potential of such an approach to compute gas-solid-liquid flow and to optimize slurry bubble columns reactors.

A potential practical application of the proposed CFD approach is to design a coal-slurry oxygen fed gasifier of interest in the FutureGen project. It is expected that such a gasifier will be much smaller and more efficient than those being tested in the FutureGen project. Such computations should hopefully clearly demonstrate the power of CFD to make design improvements.

## Nomenclature

| | |
|---|---|
| $H_k$ | enthalpy of phase k |
| $\dot{m}$ | mass transfer rate |
| $M^{jx}$ | molecular weight of the *jx*th species |
| P | phase pressure |
| $v$ | hydrodynamic velocity |
| $y'_{jx}$ | Gas mole fraction of the *jx*th species in k phase |
| $\beta_B$ | interphase momentum transfer coefficient |
| $\varepsilon$ | phase volume fraction |
| $\gamma_k$ | energy dissipation due to inelastic particle collision |
| $\Theta$ | granular temperature |
| $\rho$ | phase density |
| $\tau$ | phase stress |

## References


1. Fan, L.-S. *Gas-Liquid-Solid Fluidization Engineering*; Butterworth: Boston, MA, 1989.
2. Tarmy, B. L.; Coulaloglou, C. A. *Chem. Eng. Sci.* **1992**, *47*, 3231.
3. Parkinson, G. *Chem. Eng.* **1997**, *104*, 39.
4. Katzer R.J.; Ramage, M. P.; Sapre, A. V. *Chem. Eng. Progress* July, 2000, p 41.
5. Tullo, A. H. *Chem. Eng. News* July, 2003, p 18.
6. Stiegel, G. J. *PETC Review*; PETC office of Fossil Energy-U.S. DOE: Liquid Transportation Fuels from Coal - Part2: Indirect Liquefaction., Fall, 1991, Issue 4.
7. Heydorn, E.C.; Street, B.T.; Sarkus, T.A.; Kornosky, R.M.; Miller, C.L. *Clean Coal Technology*; DOE Topical Report: Commercial-Scale Demonstration of the Liquid Phase Methanol (LPMEOH$^{TM}$) Process., April, 1999, Number 11.

8. Ramachandran, P. A.; Chaudhari, R. V. *Three-Phase Catalytic Reactors*; Gordon and Breach Science Publishers, New York, NY, 1983.
9. Deckwer, W. D. *Bubble Column Reactors*; Wiley, New York, NY, 1992.
10. Prakash, A.; Bendale, P. G. *Design of Slurry Reactor for Indirect Liquefaction Applications*; Final Report to DOE/PETC: Viking Systems International, 1991.
11. van der Laan, G. P.; Beenackers, A. A. C. M.; Krishna, R. *Chem. Eng. Sci.* **1999**, *54*, 5013.
12. Pan, Y.; Dudukovic, M. P.; Chang, M. *AIChE J.* **2000**, *46*, 434.
13. Dudukovic, M. P. *International Conference on Multiphase Flows*, New Orleans, LA, May 27- June 1, 2001.
14. Mitra-Majumdar, D.; Farouk, B.; Shah, Y. T. *Chem. Eng. Sci.* **1997**, *52*, 4485.
15. Matonis, D.; Gidaspow, D.; Bahary, M. *AIChE J.* **2002**, *48*, 1413.
16. Nowocki, L.; Ledakowicz, S.; Bukur, D. B. *Chem. Eng. Sci.* **2001**, *56*, 1175.
17. Shah, Y. T. ; Kelkar, B. G. ; Godbole, S. P. ; Deckwer, W.-D. *AIChE J.* **1982**, *28*, 353.
18. Gidaspow, D. *Multiphase Flow and Fluidization: Continuum and Kinetic Theory Descriptions*; Academic Press, Inc., New York, NY, 1994.
19. Wu, Y.; Gidaspow, D. *Chem. Eng. Sci.* **2000**, *55*, 573.
20. Gamwo, I. K.; Gidaspow, D.; Jung, J. *Ind. & Eng. Chem. Res.* **2005**, *44*, 6393.
21. Savage, S. B.; Jeffrey, D. J. *J. Fluid Mech.* **1981**, *110*, 255.
22. Lun, C. K. K.; Savage, S. B.; Jeffrey, D. J.; Chepurniy, N. *J. Fluid Mech.* **1984**, *140*, 223.
23. Chapman, S.; Cowling, T. G. *The Mathematical Theory of Non-Uniform Gases;* Cambridge, New York, NY, 1970.
24. Gidaspow, D.; Jung, J.; Singh, R. K. *Powder Tech.* **2004**, *148*, 123.
25. Ahmadi, G.; Ma, D. *Int. J. Multiphase Flow* **1990**, *16*, 323.
26. Cao, J.; Ahmadi, G. Int. *J. Multiphase Flow* **1995**, *21*, 1203.
27. Graaf, G. H.; Winkelman, J. G. M.; Stamhuis, E. J.; Beenackers, A. A. C. M. *Chem. Eng. Sci.* **1988**, *43*, 2161.
28. Graaf, G. H.; Beenackers A. A. C. M. *Chem. Eng. and Processing* **1996**, *35*, 413.
29. Tartan, M.; Gidaspow, D. *AIChE J.* **2004**, *50*, 1760.
30. Mudde, R. F.; Lee, D. J.; Reese, J.; Fan, L. S. *AIChE J.* **1997**, *43*, 913.
31. Bahary, M. Ph.D. thesis, Illinois Institute of Technology, Chicago, IL, 1994.
32. Chen, J. J. J.; Jamialahmadi, M.; Li, S. M. *Chem. Eng. Des.* **1989**, *67*, 203.
33. Jung, J. Ph.D. thesis, Illinois Institute of Technology, Chicago, IL, 2003.
34. Schulz, H. *Applied Catalysis A: General*, **1999**, *186*, 3.
35. Grace, J. R. In *Gas Fluidization Technology;* Editor Geldart, D.,Ed.; John Wiley & Sons Ltd: New York, NY, 1986. pp 285.

36. Maretto, C.; Piccolo, V. U.S. Patent 5,827,902, 1998.
37. Clark, G. L.; Walker, D. G. U.S. Patent 6,156,809, 2000.
38. Cody, G. D.; Goldfarb, D. J.; Storch, Jr. G. V.; Norris, A. N. *Powder Tech.* **1996**, *87*, 211.
39. Chang, M.; Coulaloglou C. A. U.S. Patent 5,252,613, 1993.
40. Gamwo, I. K.; Halow, J. S.; Gidaspow, D.; Mostofi, R. *Chem. Eng. J.* **2003**, *93*, 103.
41. Hinds, W. C. *Aerosol Technology : Properties, Behavior, and Measurement of Airborne Particles;* John Wiley & Sons, New York, NY,1982.
42. Gamwo, I. K.; Gidaspow, D. *Computational and Experimental Modeling of Three-phase Slurry Bubble Column Reactor*; Annual Report to DOE: Illinois Institute of Technology, 1999.
43. Behkish, A.; Men, Z.; Inga, J.R.; Morsi, B.I. *Chem. Eng. Sc.* **2002**, *57*, 3307.
44. Vandu, C.O.; Krishna, R. *Chem. Eng. Technol.* **2003**, *26*, 778.
45. Kim, J.; Moin, P.; Moser, R. *J. Fluid Mech.* **1987**, *177*, 133.
46. Jiradilok, V.; Gidaspow, D.; Damronglerd , S.; Koves, W. J.; Mostofi, R. AIChE 2005 annual meeting CDROM preprints, Fundamentals of Fluidization session 2, 2005.

# Indexes

# Author Index

# Subject Index

**D**

E

**F**

## G



## H

## I

## Q



## R

## S

## T



## U



## V

## W



## Z